# Cabin Crew Conference 2017



Held at the 2017 Asia Pacific Airline Training Symposium
(APATS 2017)

Singapore
29 - 30 August 2017

ISBN: 978-1-5108-4925-9

**Printed from e-media with permission by:**

Curran Associates, Inc.
57 Morehouse Lane
Red Hook, NY 12571

**Some format issues inherent in the e-media version may also appear in this print version.**



Printed by Curran Associates, Inc. (2017)



Halldale Media Inc.
735 Primera Blvd.
Suite 220
Lake Mary, Florida 32746
USA

Phone: (407) 322-5605
Fax: (407) 322-5604

www.halldale.com

**Additional copies of this publication are available from:**

Curran Associates, Inc.
57 Morehouse Lane
Red Hook, NY 12571 USA
Phone: 845-758-0400
Fax: 845-758-2633
Email: curran@proceedings.com
Web: www.proceedings.com

# TABLE OF CONTENTS

apats2017 SINGAPORE

AA Nova
Training for Excellence

# Refocusing Cabin Crew Training

Aye Aye Naing

Trevor Jensen

AA Nova
Training for Excellence

# SOPs Training

AA Nova
Training for Excellence

# Crew Resource Management

2017

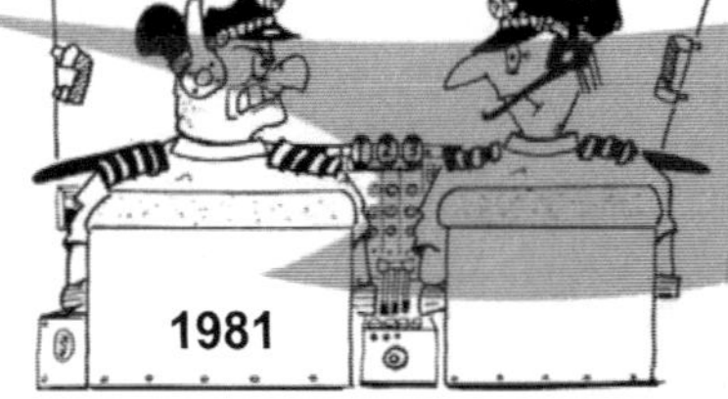


Flight attendants are probably the most obvious of the groups other than pilots who may profit from CRM training.

(FAA AC 120-51E)

AA Nova
Training for Excellence

## Has CRM really been integrated?

AA Nova
Training for Excellence

## Let's be honest….

# CRM is Cockpit Centric

AA Nova
Training for Excellence

## And 9/11 hasn't helped

AA Nova
Training for Excellence

## The winds of change are upon us

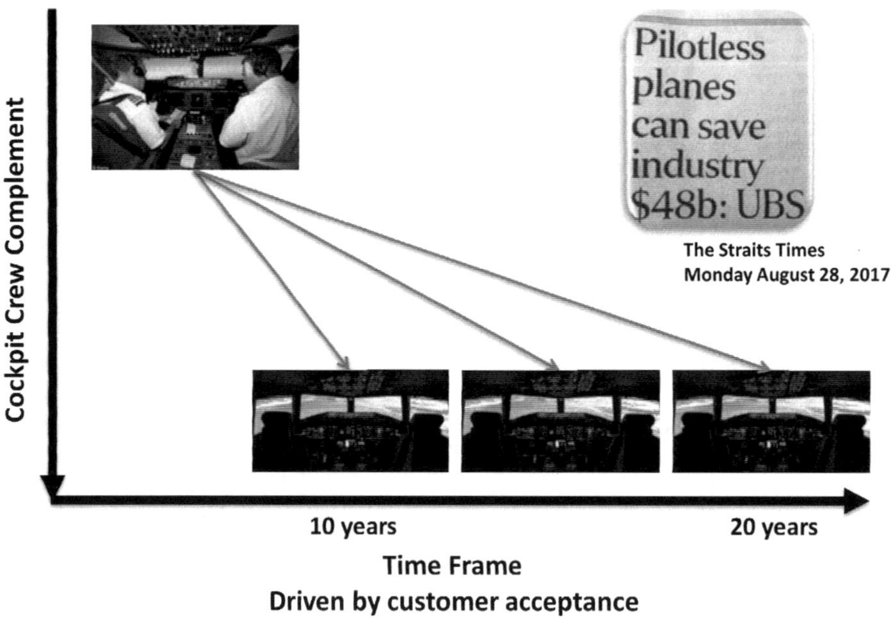


AA Nova
Training for Excellence

## What is the future cabin crew role?

AA Nova
Training for Excellence

## Should we develop a roadmap?

e.g. will current selection criteria be relevant?

e.g how will we motivate and train

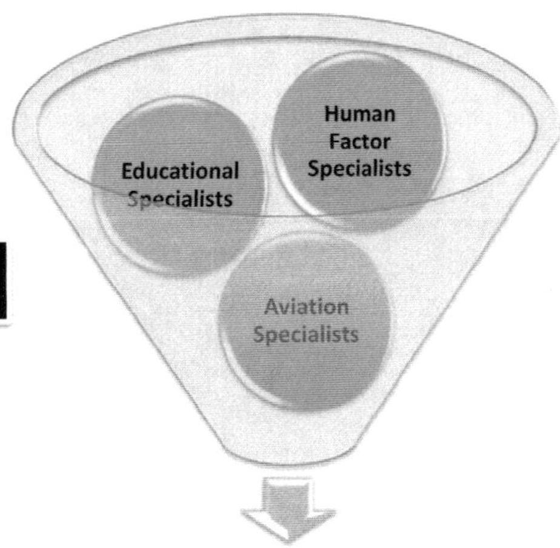


**Define the role**
**The attributes**
**The training**

AA Nova
Training for Excellence

## Authority of the CC in Charge

**Basis of Command Authority – ICAO Annex**

The Pilot in Command is responsible for the safety and disposition of the aircraft, crew and cargo.

The Operator must nominate a PIC.

**Is it time for the Authority of the Cabin Crew in Charge to be Legislated?**

The Cabin Crew in Charge is responsible to the Pilot in Command for the safety welfare of all passengers and crew under their direction.

The Operator must nominate a Cabin Crew in Charge.

AA Nova
Training for Excellence

## Start to redress working relationships

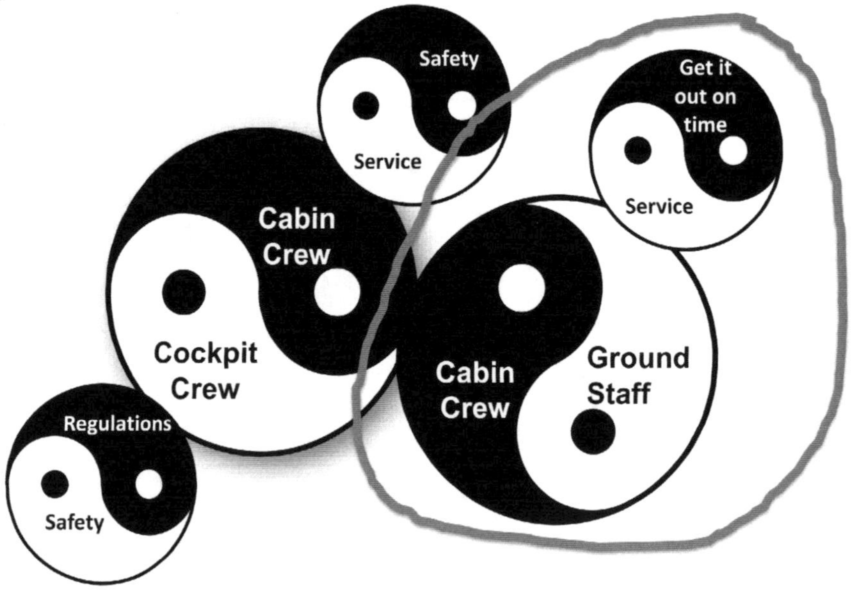


AA Nova
Training for Excellence

## Is the impact on the cabin understood?

AA Nova
Training for Excellence

AA Nova
Training for Excellence

apats2017
SINGAPORE

AA Nova
Training for Excellence

Thank you

and we look forward to hearing your thoughts

apats2017
SINGAPORE

SUSAN RICE EMAIL:TAASUSAN@GMAIL.COM TUESDAY 29 AUGUST CABIN SAFETY CONFERENCE 1

apats2017
SINGAPORE

# *How do 'we' develop & maintain competent cabin crew?*

*Susan Rice – Aviation Consultant/Cabin Safety Specialist*
*Email: taasusan@gmail.com*
*Tuesday 29 August 2017*

apats2017 SINGAPORE

☺

# *Do you like to travel and love meeting people?*

*Really? ......*

SUSAN RICE EMAIL:TAASUSAN@GMAIL.COM TUESDAY 29 AUGUST CABIN SAFETY CONFERENCE 3

apats2017 SINGAPORE

# *Do you possess a high standard of customer service skills?*

*Really?*

4

SUSAN RICE EMAIL:TAASUSAN@GMAIL.COM TUESDAY 29 AUGUST CABIN SAFETY CONFERENCE

apats2017 SINGAPORE

**Pre Application**

What opportunities are available to applicants that provide a ***'REALISTIC' & ACCURATE'*** understanding of the role & responsibilities?

SUSAN RICE EMAIL:TAASUSAN@GMAIL.COM TUESDAY 29 AUGUST CABIN SAFETY CONFERENCE 5

apats2017 SINGAPORE

SUSAN RICE EMAIL:TAASUSAN@GMAIL.COM TUESDAY 29 AUGUST CABIN SAFETY CONFERENCE 6

Many Cabin Crew training organisations have evolved in recent years.
They have undertaken that they will provide an applicant with the potential and essential criteria to enable successful recruitment into an airline.



## ***RECRUITMENT***



✓**Criteria** – What qualities and skills are considered necessary to ensure the ***most suitable*** applicant is recruited as a member of Cabin Crew?

- Where does an airline operator look to seek the most contemporary data and information as it relates to optimum acquired and inherent skills?

apats2017 SINGAPORE

## Recruitment

✓ How much of the *'suitability criteria'* depends upon culture : ie the individual internal airline culture and/or individual nationality/ethnicity?

✓ Regulatory requirements:

- ❑ Is the applicant likely to possess the *physical* attributes required?
- ❑ Is it perceived the applicant has the necessary interpersonal skills to manage an abnormal situation



apats2017 SINGAPORE

## *INITIAL TRAINING*

✓ **Standards to be implemented: based upon what criteria?**

Regulatory requirements and subsequent approval of theoretical and practical training

➢ ***Ensuring realism and accuracy of all equipment, aircraft systems and the cabin environment is imperative to ensure meaningful & successful practical training!***

apats2017
SINGAPORE

*"Theoretical comprehension is fundamental*

*HOWEVER*

*Practical proficiency is a potentially life saving attribute"*

SUSAN RICE - TAASUSAY@GMAIL.COM TUESDAY 29 AUGUST 2017 CABIN CREW CONFERENCE

12

apats2017
SINGAPORE
SUSAN RICE EMAIL:TAASUSAN@GMAIL.COM TUESDAY 29 AUGUST CABIN SAFETY CONFERENCE
13

apats2017
SINGAPORE
SUSAN RICE - TAASUSAN@GMAIL.COM TUESDAY 29 AUGUST 2017 CABIN CREW CONFERENCE
14

apats2017 SINGAPORE

SUSAN RICE - TAASUSAN@GMAIL.COM TUESDAY 29 AUGUST 2017 CABIN CREW CONFERENCE 15

apats2017 SINGAPORE

SUSAN RICE EMAIL:TAASUSAN@GMAIL.COM TUESDAY 29 AUGUST CABIN SAFETY CONFERENCE 16

SUSAN RICE EMAIL:TAASUSAN@GMAIL.COM TUESDAY 29 AUGUST CABIN SAFETY CONFERENCE 17

## *INITIAL TRAINING*

✓Some <u>external/independent</u> training organisations work in close conjunction with an airline operator to deliver a 'Total Package'!

☺ At completion, a fully trained and regulatory approved member of Cabin Crew

SUSAN RICE EMAIL:TAASUSAN@GMAIL.COM TUESDAY 29 AUGUST CABIN SAFETY CONFERENCE 18

apats2017 SINGAPORE

apats2017 SINGAPORE

## *INITIAL TRAINING*

✓ **Possible realistic expectations of a qualified member of Cabin Crew:** *(but not limited to)*

- ❑ *Understanding of safety role & responsibilities*
- ❑ *Ability to demonstrate practical competence*
- ❑ *Understanding of operator regulatory obligations*
- ❑ *Ability to implement learned skills*
- ❑ *Understanding of State regulatory structure*
- ❑ *Understanding of own personal regulatory obligations*
- ❑ *Be in possession of emotional intelligence necessary*
- ❑ *Interpersonal and communication skills*

SUSAN RICE - TAASUSAN@GMAIL.COM TUESDAY 29 AUGUST 2017 CABIN CREW CONFERENCE 21

apats2017 SINGAPORE

## Recurrent Training

✓ The greatest challenge is to ensure Cabin Crew, who have graduated from their initial training, are able to maintain a level of competence equal to and then above those initial standards!

❑ **How is that commitment met and managed?**

SUSAN RICE EMAIL:TAASUSAN@GMAIL.COM TUESDAY 29 AUGUST CABIN SAFETY CONFERENCE 22

apats2017 SINGAPORE

✓ Recurrent, practical training and additional relevant human factors upskilling would prove to be of great benefit to an individual member of Cabin Crew

- Scenario based exercises utilising both practical and interpersonal skills – based upon factual incidents/accidents that have occurred

SUSAN RICE EMAIL:TAASUSAN@GMAIL.COM TUESDAY 29 AUGUST CABIN SAFETY CONFERENCE 23

apats2017 SINGAPORE

- ***Is regulatory compliance with current, minimum requirements sufficient?***
- ***ICAO has now developed documentation that will set the standards for States to ensure that in many years to come 'we' will continue to produce competent current and future Cabin Crew!***

SUSAN RICE EMAIL:TAASUSAN@GMAIL.COM TUESDAY 29 AUGUST CABIN SAFETY CONFERENCE 24

apats2017
SINGAPORE

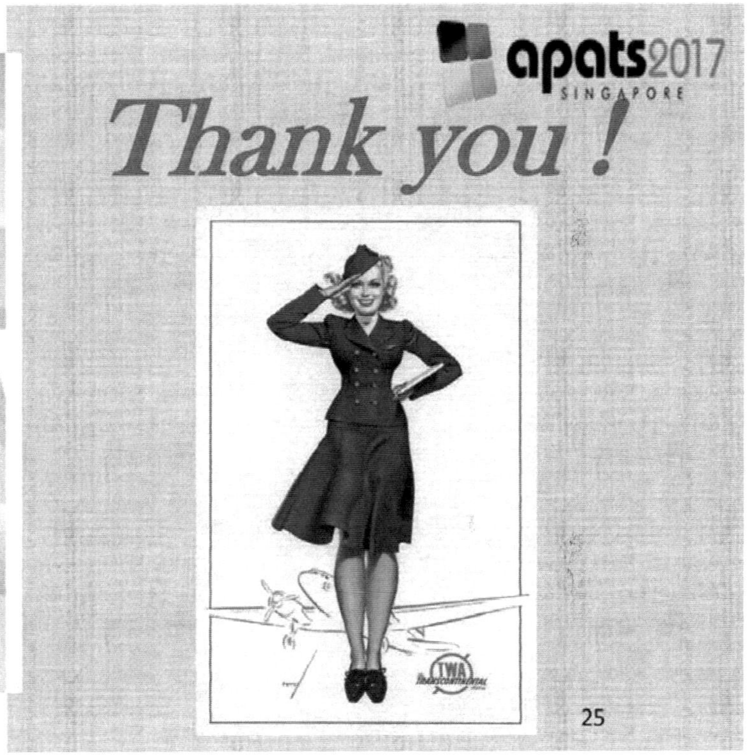
apats2017
SINGAPORE
Thank you !
25

Simple Interface

## A Complex Interface

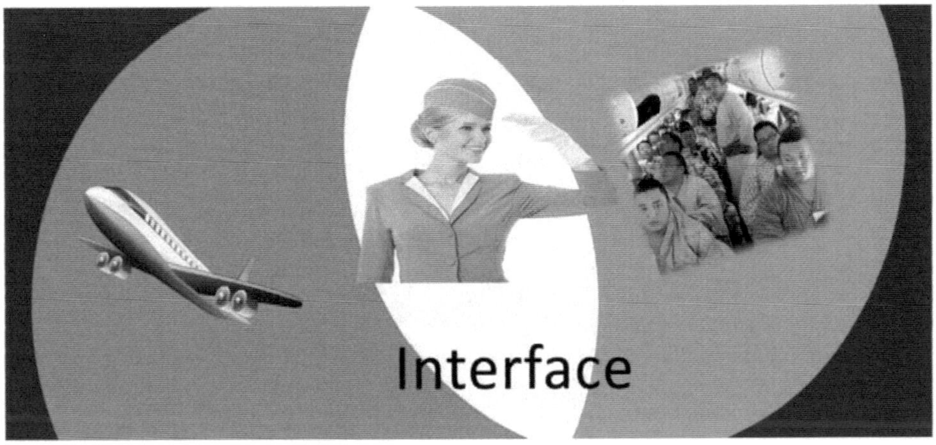


## The Challenge

Redefine Standard Practices in a Dynamic Environment

## Improving the Interface- (Upgrading Cabin Crew IOS)

- Empowering Cabin Crew to be more dynamic
- Communication with a diverse group
- Assertiveness is usually misconstrued
- Increasing empathy and reducing judgments and biases
- Increase comfort

## Perhaps one day…

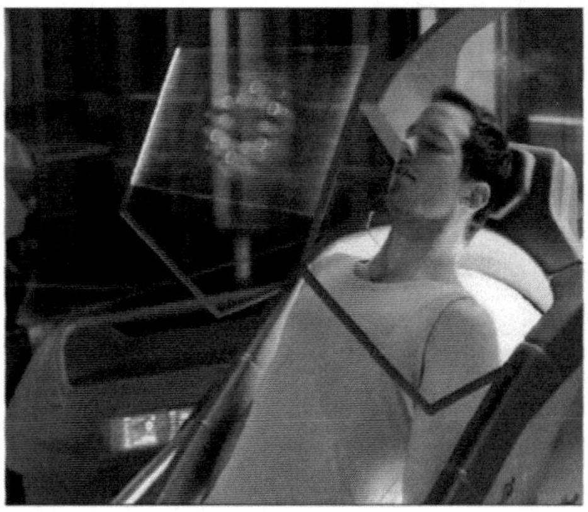

AEROFLOT Russian Airlines

JEDIUM EXPANDING INTELLIGENCE

## Using Dynamic Virtual Environments for Cabin Crew Training
(case study)

**Olga Chashina** – Aeroflot, Information Systems Department
**Vitaly Chashin** – Aeroflot, Information Systems Department
**Sergey Kudryavtsev** – Jedium Inc., CEO

## About Aeroflot

- The leading aviation company in Russian Federation
- Member of SkyTeam Airline Alliance
- **198 airplanes** in total, including *B777, B737, A330, A321, A320* and *SSJ-100*

AEROFLOT Russian Airlines



JEDIUM EXPANDING INTELLIGENCE

## Education in Aeroflot

- Two main training departments: **Internal** and the **Aviation School**
- Blended learning (LCMS, Adobe Connect, classroom)
- A number of internal e-learning applications (simulations, speech training, etc.)
- Cabin crew training:
  - **7800 people (!)**
  - Yearly: **season training, refresher training**
  - Training for aircraft type: **refresher training, recurrent training**
  - **Senior cabin crew training**
  - Classroom training: **CRM, safety equipment & procedures, first aid, customer service, language training**

AEROFLOT Russian Airlines 3 JEDIUM EXPANDING INTELLIGENCE

## Problems

**Main problems:**

- Lack of teachers/instructors
- Current e-learning process takes a lot of time

**Possible solutions:**

- **Extended adaptive course/personal learning trajectory**
  - *Aeroflot already uses "continuous education" system, which includes an adaptive-course capability*
- **VR**
  - *Aeroflot has been working on VR training since 2010*
  - *VR has been already used in cabin crew language training (in some degree)*

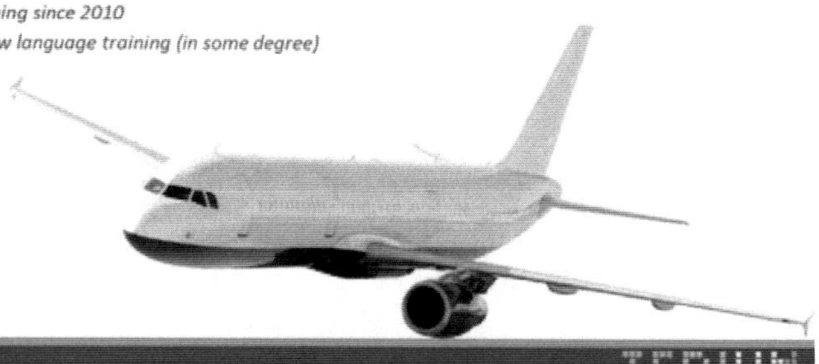

AEROFLOT Russian Airlines 4 JEDIUM EXPANDING INTELLIGENCE

## Virtual Environment

Types:

- *360 Video?*
- Mobile VR
- PC VR

Already widely used in training. One of the future trends in education

- Equipment training
- Situation training

There are no typical design patterns for VR in education now

AEROFLOT Russian Airlines 5 JEDIUM EXPANDING INTELLIGENCE

## Main Problems of VR in Education

**Cost of hardware + Inflexibility in buying options**

- New hardware is released. As a result, costs go down
- VR vendors are getting focused on B2B

**Hard to develop**

- SDKs are not mature
- The vendors require optimized experiences
- Development requires specific programming skills

**Typically requires a new project for every e-learning experience**

- High Total Cost of Ownership (TCO)
- Hard to create an ecosystem: course & version management are difficult

**Hard to integrate into an existing infrastructure**

- Especially in terms of learning statistics

AEROFLOT Russian Airlines 6 JEDIUM EXPANDING INTELLIGENCE

## Dynamic Virtual Environment (DVE)

Technically: a multi-user virtual environment, where users can create new content

- Not a new idea, but it's not widely used (difficult to implement)

From the e-learning point of view:

- Lots of new possibilities
- Requires a carefully desigred workflow
- Still has some problems with integrations

## Collaborative Modes in DVE

**Single-user**

- Can replace a number of typical e-learning tests with a reach content
- Requires some kind of design toolset (and workflow)
- Can use traditional integrations with LMS (like usual tests)

**Virtual classroom**

- Works as an addition to webinars. In some cases, it solves webinar-related problems (3D content, etc.)
- Easy to implement (if you have a DVE)

**"Thematic" classroom or a virtual demonstration**

- Works better than default classroom or webinar
- Requires design tools
- Has an ability to use bots

**Role-playing games and simulations**

- Efficient training
- Has an ability to use bots
- Requires design tools and, specifically, a scenario designer
- Requires a different way of an integration with existing infrastructure

# Design tools for DVE

**Typical solution for DVE:**

- The user with required access level can change the environment straight in place while being there
  - *Works well only for small changes*
  - *Hard to set up some kind of workflow*
  - *Problems with User Experience, User Interface*

**Two types of design objects:**

1. 3D objects and behaviors
   - An object can (and should) be separated from the scenario
   - "Black box": has some input interactions (like touch, pick up, etc.) and outputs (like animation, color change, etc.)
   - Must be "multi-user" by design
   - **Can be reused**
2. Scenarios
   - Hard to design "in-world"
   - Describes some kind of situation in terms of interaction between users and objects
   - May have a number of outcomes
   - There are already a number of languages and tools for them (Twine, Scrivener, etc.)

**WE SEPARATE THE DESIGN OF OBJECTS AND THEIR BEHAVIORS FROM DESIGN OF SCENARIOS**



# Integrations and xAPI

**Integrations are one of the key problems for VR experiences**

- Especially in case of role-playing games, which are different by the nature from typical quizzes and tests

**xAPI – designed for serious games**

- "noun-verb-object" statements
- **Could be used on object and behavior levels**
- **Could be used on a scenario level**

## Personal Big Data

**Using xAPI on scenario level allows easy integrations with current LMS/LCMS**

- Like in the standard quizzes, we save the outcomes of different scenarios

**Using xAPI on an object level**

- An ability to record every interaction
- **Personal Big Data**

**Typical size of one multi-user scenario "walktrough" is enough to use ML and data mining**

- Especially in case of "mixing" multiple scenarios in one experience, and it's one of the key abilities of DVE

AEROFLOT Russian Airlines 11 JEDIUM EXPANDING INTELLIGENCE

## Jedium

The project is based on *Jedium Inc.* software

Entity-based approach (every object in the DVE is an independent entity)

Highly scalable:

- 200 and more users in one virtual environment
- In fact, limited only by client-side hardware. That's the same for objects

Has an ability to code complex behaviors of objects (full-scale C, .NET)

Works with all major desktop VR solutions

AEROFLOT Russian Airlines 12 JEDIUM EXPANDING INTELLIGENCE

## Adaptation for Aeroflot

Scenario editor

- The platform already had inworld editor, but scenario editor was required to create a stable workflow

Bots

- Aeroflot scenarios using bots a lot, so this ability was implemented for the platform

xAPI storage and ability to capture per-scenario and per-object data

- We used open-source implementation of storage
- Made every object on the platform "trackable", i.e. saving every interaction with it in xAPI format

ElasticSearch and Kibana for xAPI data visualization and analysis



## Conclusion

Intermediate project results show the efficiency of VR training

VR could be used in a various types of training

It's possible to set up a workflow for training development and management by using the specific implementation of VR training (DVE)

It's easy to develop an integration of VR training with existing infrastructure by using the specific implementation of VR training (DVE)

Using VR in DVE implementation leads to Personal Big Data collection and analysis

## Video Presentation: VR

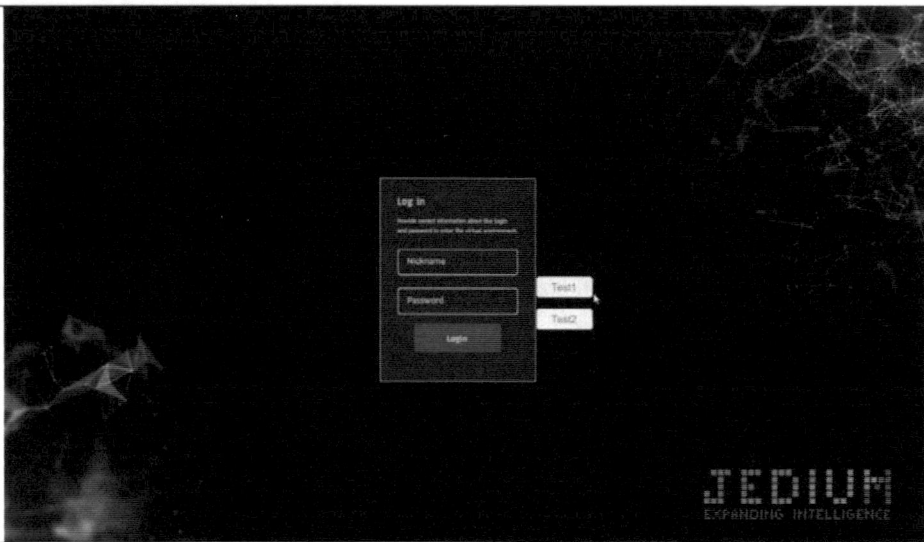


AEROFLOT Russian Airlines 15 JEDIUM EXPANDING INTELLIGENCE

## Video Presentation: Non-VR

AEROFLOT Russian Airlines 16 JEDIUM EXPANDING INTELLIGENCE

Olga Chashina
**Aeroflot – Russian Airlines**
ochashina@aeroflot.ru
www.aeroflot.ru

Vitaly Chashin
**Aeroflot – Russian Airlines**
www.aeroflot.ru

Sergey Kudryavtsev
**Jedium Inc., CEO**
sk@jedium.com
*www.jedium.com*
*www.fb.com/jedium*

# Thank you!

EYES
WIDE
SHUT:
ARE WE DOING ENOUGH TO COMBAT ONE OF
THE MOST SERIOUS HUMAN RIGHTS VIOLATIONS IN TODAY'S WORLD?
AVION LLC
TRAINING

Arrow

Vice

Dog

Top-Hat

House

Can you see the word FLY?

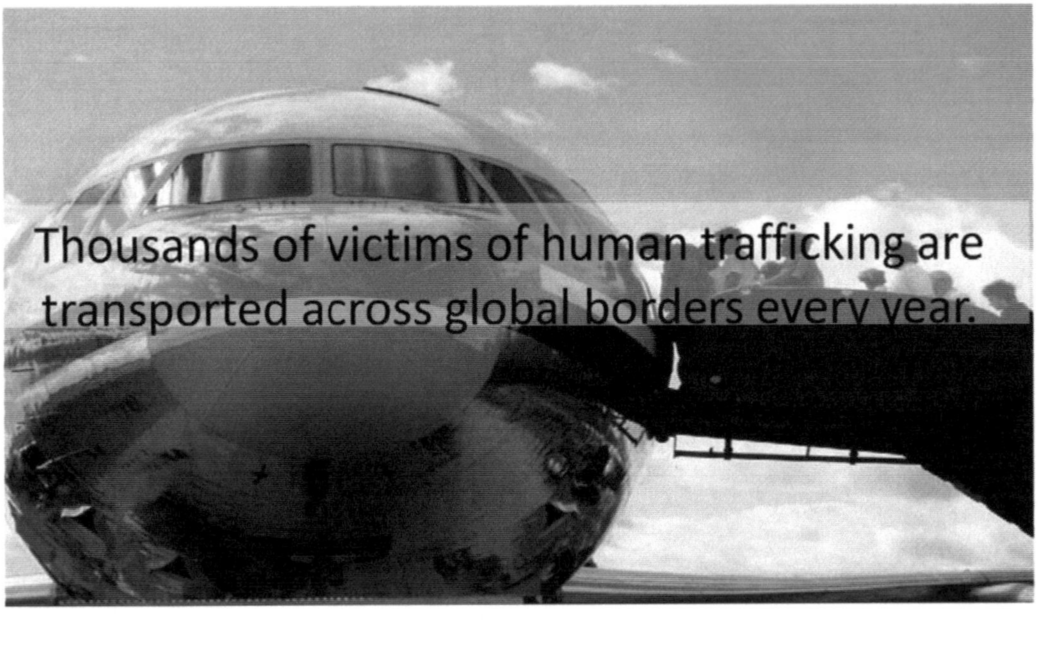

# ***'TRAFFICKING'***

*Trafficking is always about the* ***exploitation*** *of* ***vulnerabilities***

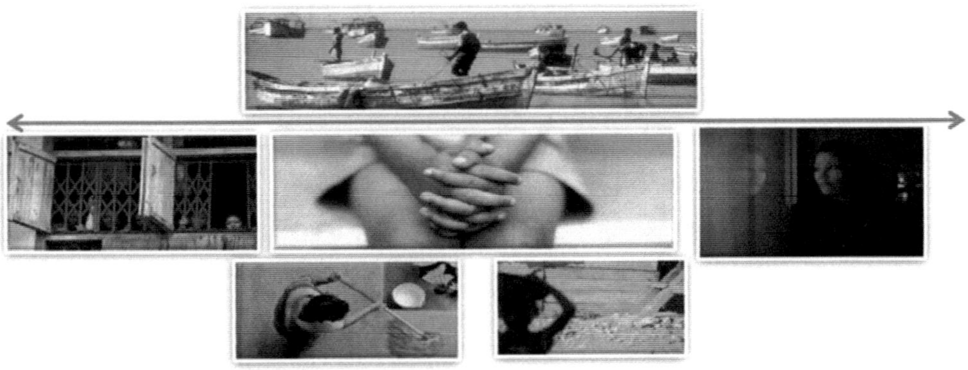

Marisol's story

# How Big is the Problem?

20-30 Million Slaves world-wide

600,000 -800,000 People a Year: world-wide (3,287 people sold, kidnapped or forced into slavery daily, 136 an hour)

80% Women & Girls –50% Are Minors

80% Sexual Exploitation

19% Labor Exploitation

UNICEF estimates 1.2 million children are trafficked every year – LGBTQ kids are especially at risk.

- Human trafficking has become the fastest growing profitable crime in the world, increasing at a rate that will soon surpass even the drug trade.
- Is estimated to make **$150 billion** in illegal profits annually.

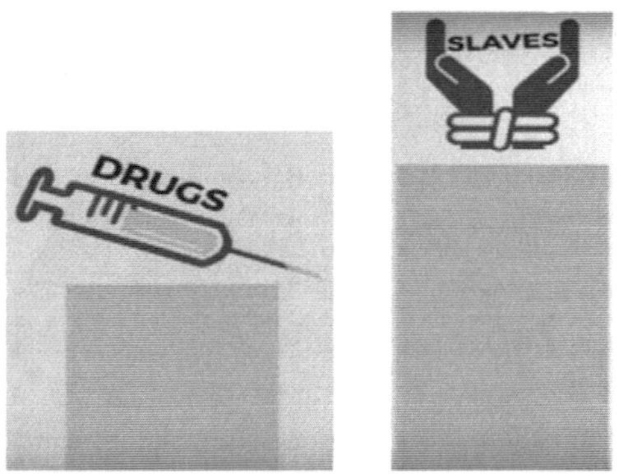

# Human Trafficking:

- Takes many forms, and mean many things, but the purpose of human trafficking is **ALWAYS exploitation.**
- Forced Prostitution/ Slaver Boats/ Domestic Servitude / Economic Exploitation/ Forced Labor/ Child Soldiers/ Organ donors

# Why WE must be part of the solution

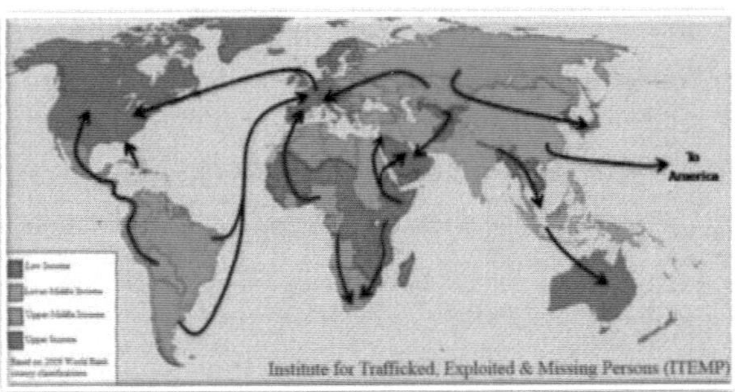


- Nearly every country is involved – either as a country of origin, destination or transit.
- Perpetrators of these heinous crimes use airlines to transport victims.
- Airline employees are in a unique position to sever a critical link by identifying and reporting trafficking victims.

**Affluent countries drive demand - victims are trafficked from poorer to wealthier countries.**

**Flight Attendant Sheila Fedrick had an uneasy feeling about a teenage girl on her flight**

**Sheila noticed an odd couple:**

- The teenager, sitting on the window seat, was "disheveled with greasy blond hair and looked like she had been through pure hell."
- Was accompanied by an older man who was very well dressed.

***This contrast between the two told her something was not right***

- She tried to speak to the two passengers but the man would not let her talk.
- The girl would not talk or look at her and only the man would respond
- He became defensive and tried to isolate the girl.
- When the girl indicated that she wanted to use the bathroom the man followed her and stood outside the door.

# Training is a Critical Key

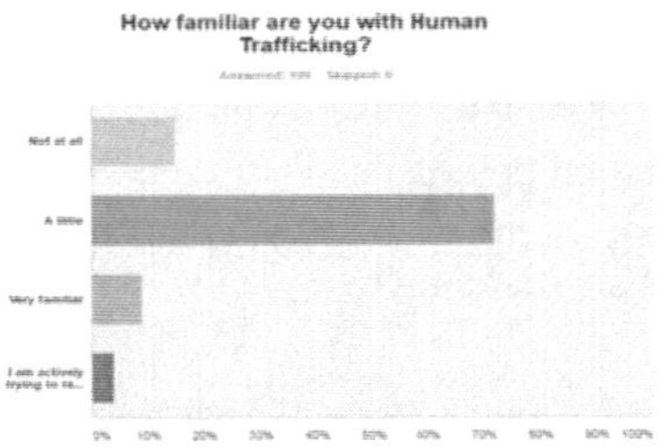


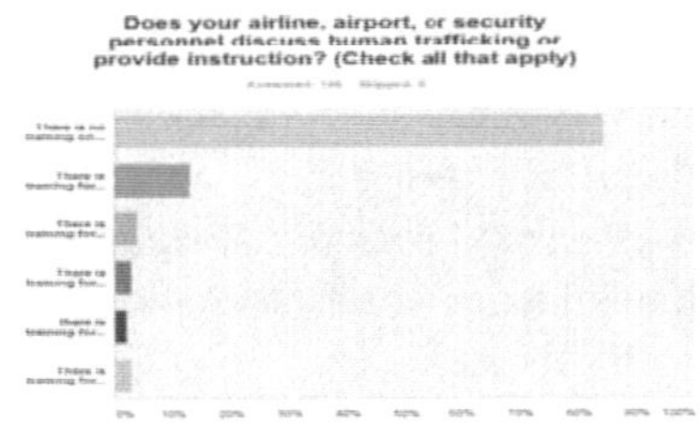


- Survey of 199 aviation respondents – 2016 (5 years later)
  - Almost 72% surveyed were a *"little"* familiar with the problem
  - 84% said their airline did not provide training

*" …I have been flying as a flight attendant for 33 years and there have been a few times before I was fully aware of the trafficking trade that I wish I could turn the clock back and do something different…"*

*"…I didn't know how or where to go to report a concern…"*

*"…I spoke to the captain. He was not willing to do anything so no further action was taken…"*

# We have work to do –

## WOULD YOU RECOGNIZE THE SIGNS?

"I thought I was going to North Carolina."
(Final destination was unclear.)

Two teenagers, visibly upset, wouldn't establish eye contact. (Final destination was unclear.)

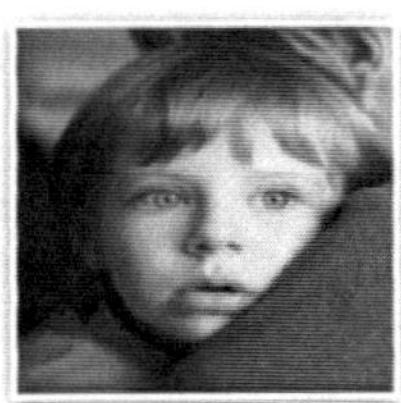

Man did not appear to be the father. (The child appeared to be drugged.)

# Some Other Possible Indicators

- Have few or none of the usual personal items when checking in or boarding a flight.
- Do not appear to know where they are going.
- Are unusually submissive to the person(s) accompanying them.
- Avoid eye contact or are watchful to the point of paranoia.
- Appear to be accompanied by someone claiming to be the parent or guardian who is in fact not a relation.

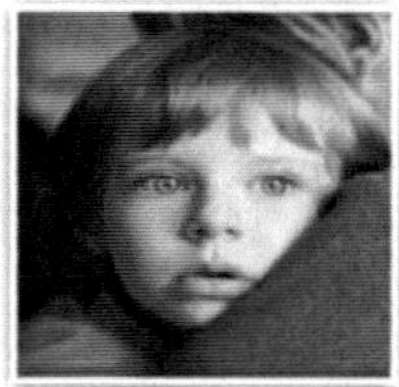

- Do not have control of their travel documents.
- May show signs of having been denied food, water, sleep, or medical care.

# Some Other Possible Indicators

- May exhibit signs of physical abuse, being drugged or unusual tattoos.
- Speak of a "modeling" job or something similar without knowing who will be meeting them upon Arrival.

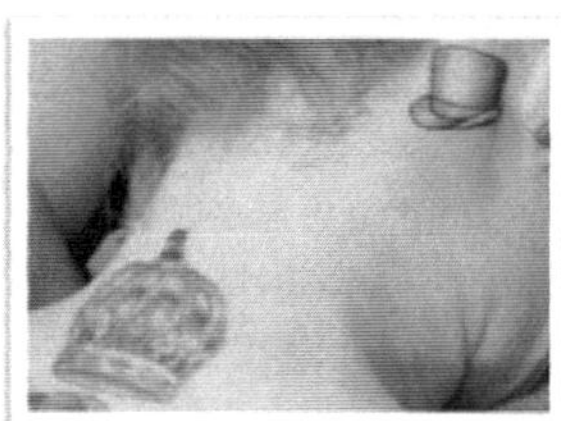

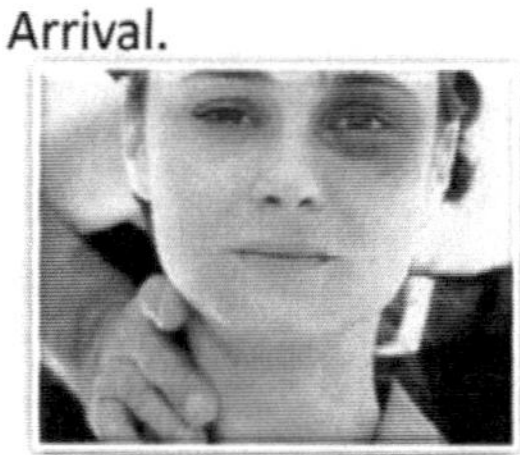

- All of these are just *indicators* – not proof, however a combination of signs should be a red flag.

BLUE CAMPAIGN
One Voice. One Mission. End Human Trafficking.™

*voluntary mechanism* to identify potential human trafficking victims and notify authorities.

ICAO SAFETY

ICAO Cabin Safety Group (ICSG)

*Requires initial and annual training*

U.S. Department of Transportation
Federal Aviation Administration

UNODC
United Nations Office on Drugs and Crime

Airline Ambassadors International

ASSOCIATION OF FLIGHT ATTENDANTS-CWA, AFL-CIO

Transport Canada

Urging flight attendants to take training – but at *discretion of airlines*

Airline Ambassadors International

"Tip Line " APP will enable law enforcement officials to receive data-rich information.

- Camera
- Voice recorder
- Hotline numbers
- Links to Tip news

AAI put a survey out to flight Attendants
- of the 374 responses
trafficking has been reported 280 times!

| | |
|---|---|
| Air Canada | Federal Aviation Administration |
| Air Canada Rouge | Flight Attendant School |
| Air China | International Air Transport Association |
| Air Transat | International Civil Aviation Organization |
| Airbus | International Society of Air Safety Investigators |
| Alitalia | International Transport Workers' Federation |
| All Nippon Airways | Jamaica Civil Aviation Authority |
| Avianca | Jet2.com |
| Boeing Company | Jetstar Airways |
| Bombardier Aerospace | LATAM Airlines |
| Central American Agency for Aviation Safety | Lufthansa Flight Training |
| Civil Aviation Administration of China | National Civil Aviation Agency of Brazil. |
| Civil Aviation Authority of Singapore | National Transportation Safety Board |
| Civil Aviation Bureau of Japan | Pegasus Airlines |
| Civil Aviation Safety Authority | Provincial Airlines |
| Civil Aviation University of China | Saudia Airlines |
| Condor Flugdienst | Sky Regional |
| Deutsche Lufthansa | Singapore Airlines |
| DINACIA Uruguay Civil Aviation Authority | South Africa Civil Aviation Authority |
| Directorate General of Civil Aviation of Turkey | Transport Canada Civil Aviation |
| Embraer | Turkish Airlines |
| Emirates Airline | UAE General Civil Aviation Authority |
| European Aviation Safety Agency | United Airlines |
| Federal Air Transport Agency (Rosaviatsia) | Vueling Airlines |

# But we all can do more

- With the proper training you can play a critical role because you are uniquely positioned to observe passengers and notice unusual behavior.

- You are all together for a longer period of time.

- <u>The 1[ST] step</u> is: Recognizing the signs

27

# What else can you do?

- Be observant especially women and young children traveling alone. Look for unusual behavior or signs.

- Note who is traveling with them and if someone controls their movement.

- *Initiate pleasant, non-threatening conversation.*

28

# Know your country's TIP line numbers

Work with your crew to establish a protocol for reporting suspected trafficking incidents on board.

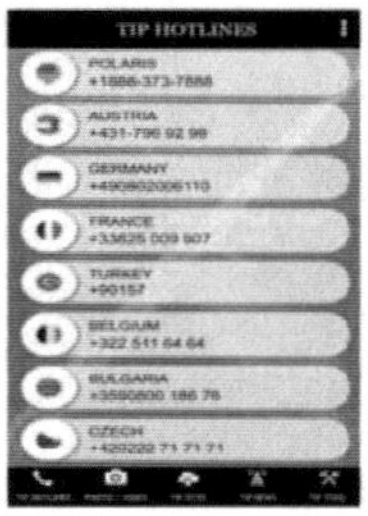

- In the US *DHS Tip Line* **866.347.2423**
as a back-up call airport police or **911**

# Just a reminder...

- Do not confront.
- Do not try to rescue.
- Do not display alarm or unusual concern.
- Do not endanger your self or others.

# YOU CAN MAKE A DIFFERENCE!

Keep your eyes WIDE OPEN – look beneath the surface

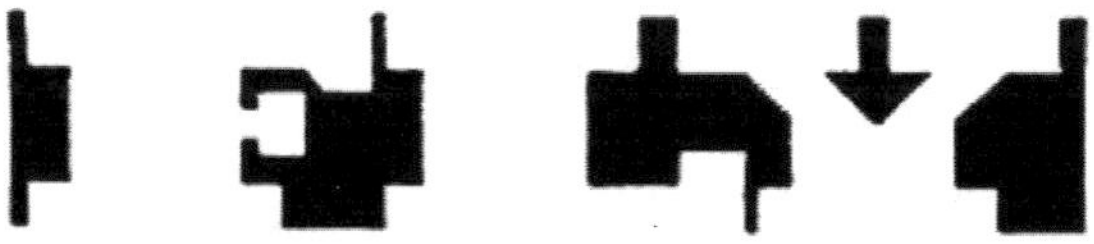

- Learn to Recognize the signs of trafficking
- Know how/where to report a suspected incident
- Urge your airlines to get involved

**...because you might be the last person with a chance to help save a life**

AVION LLC
TRAINING

# Thank You

Sherry Saehlenou
www.aviontraining.com

AEROTEAM

LEARNING EFFECT?

TRAINING TEAMWORK?

# LEARNING GAMES

SIMULATIONS?

NOTECHS?

**FOR TRAINING AND ASSESSMENT OF NOTECHs IN THE CLASSROOM ENVIRONMENT**

**BY**

**SØREN SEINDAL AGNER**

**AEROTEAM**





AEROTEAM

# INTRODUCTION

- **Søren Seindal Agner, AEROTEAM**
  - Tour Guide, Cabin Crew, Operations
  - Pilot
    - Danish Air Transport (DAT)
    - Scandinavian Airlines (SAS)
  - Ground Instructor
  - CRMT/E
  - MCC Instructor
  - Training Manager
  - Member of The Danish Flight Safety Council
    - Human Factors in Civil Aviation

AEROTEAM

## What is CRM?

*CRM is the effective utilisation of all available resources (e.g. crew members, aircraft systems, supporting facilities and persons) to achieve safe and efficient operation.*

*The objective of CRM is to enhance the communication and management skills of the flight crew member concerned. Emphasis is placed on the non-technical knowledge, skills and attitude of flight crew performance*

***EASA GM1 ORO.FC.115***





AEROTEAM

## What is CRM?

*CRM is the effective utilisation of all available resources (e.g. crew members, aircraft systems, supporting facilities and persons) to achieve safe and efficient operation.*

*The objective of CRM is to enhance the communication and management skills of the crew member, as well as the importance of effective coordination and two-way communication between all crew members.*

***EASA GM1 ORO.CC.115(e)***

AEROTEAM

# Error Management Strategies

- **Understanding the nature and extent of error or risk**
- **Changing the conditions that induce the error**
- **Determining the behaviours that prevent or mitigate error**

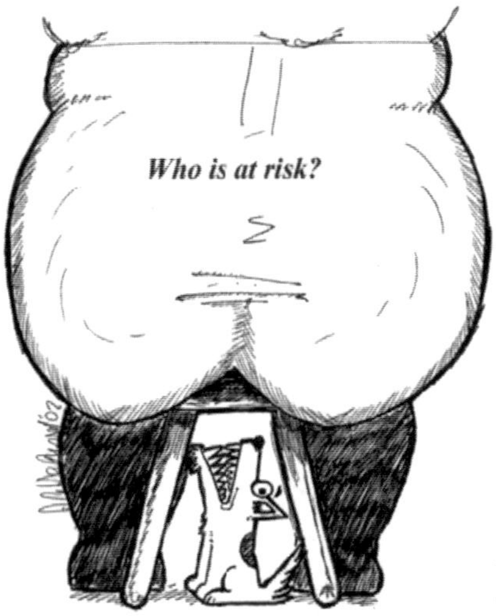




TS Presentation
AEROTEAM
02-170824/SSA

AEROTEAM

# ABILITY

SYNERGY OF YOUR ATTITUDE, KNOWLEDGE AND SKILLS

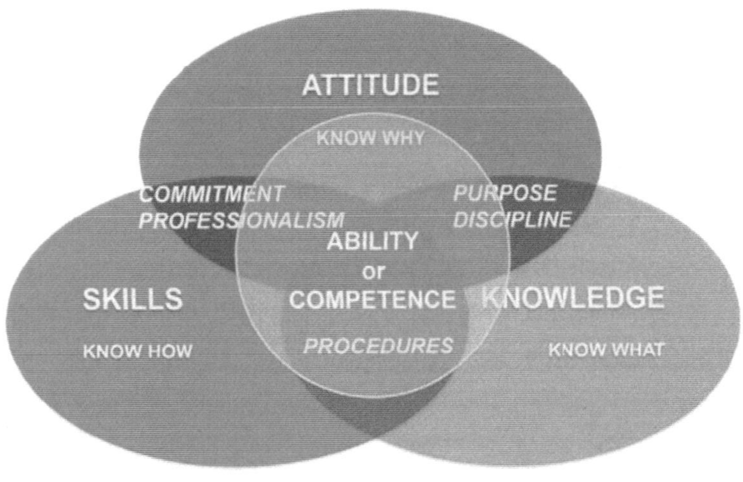




APATS Presentation
AEROTEAM
VER02-170824/SSA

AEROTEAM

# LEARNING GAMES

ANALOGUE OR DIGITAL

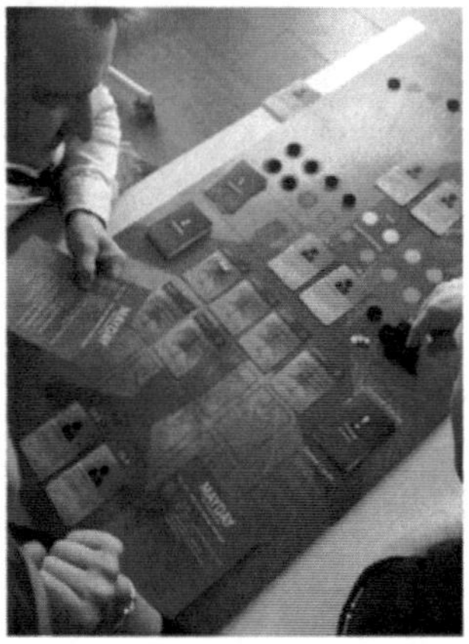

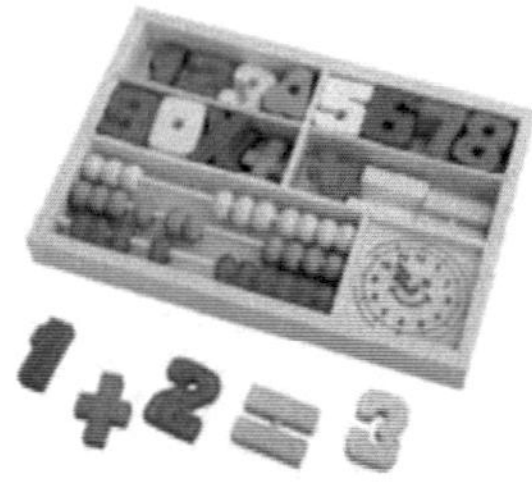






AEROTEAM

# Game-based learning

- Game-based learning is a teaching method that utilizes analogue or digital games in instructions. It involves both instructional and participatory acts:
  - **Instructional Act –** Instructors reference self-contained systems in analogue or digital games to teach concepts, theories, system-thinking, and design-thinking.
  - **Participatory Act** – learners play games and create game systems for experiential learning.

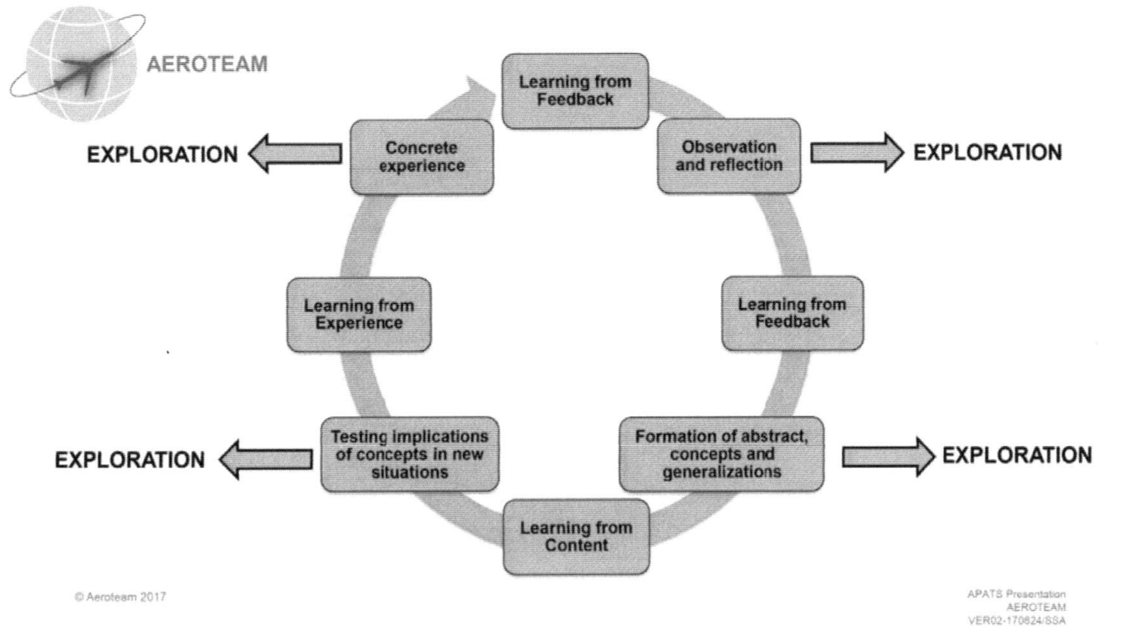
AEROTEAM
Learning from Feedback
Concrete experience
Observation and reflection
EXPLORATION
EXPLORATION
Learning from Experience
Learning from Feedback
Testing implications of concepts in new situations
Formation of abstract, concepts and generalizations
EXPLORATION
EXPLORATION
Learning from Content
© Aeroteam 2017
APATS Presentation
AEROTEAM
VER02-170824/SSA

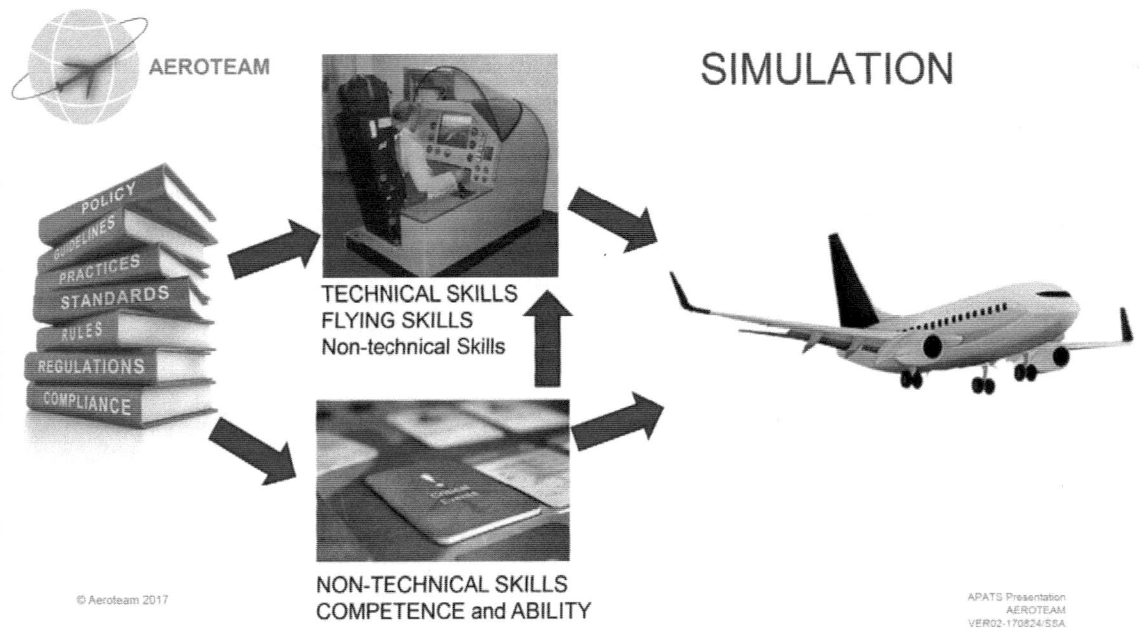
AEROTEAM
SIMULATION
POLICY
GUIDELINES
PRACTICES
STANDARDS
RULES
REGULATIONS
COMPLIANCE
TECHNICAL SKILLS
FLYING SKILLS
Non-technical Skills
NON-TECHNICAL SKILLS
COMPETENCE and ABILITY
© Aeroteam 2017
APATS Presentation
AEROTEAM
VER02-170824/SSA

AEROTEAM

# SIMULATION

- Learning by Doing or
  - Learning by Simulation
- Learning is an internal process
- Simulation creates a situation where ideas, decisions, or actions can be tried out
- It is "OK" to make mistakes
- Consequences are visible
- No actual risk

APATS Presentation
AEROTEAM
VER02-170624/SSA

AEROTEAM

# SOCIAL LEARNING

- Safe to be challenged
- Mistakes can be evaluated
- Individual learning → in a dynamic social environment
- Group decisions and evaluation (another set of eyes)
- Enhancement of knowledge and experience (by sharing)
- Platform for future cooperation

APATS Presentation
AEROTEAM
VER02-170624/SSA

AEROTEAM

## Overview of Complex Connections

- Simulation of complex situations and events
- Visual presentation of decisions, consequences and actions
- Easier to see the full picture

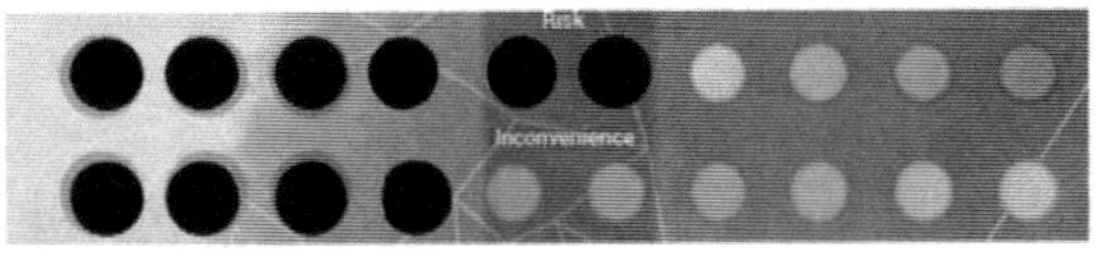

AEROTEAM

## ENGAGEMENT and MOTIVATION

- Playing games is generally fun.
- Learning while playing makes it fun to learn
- Focus must be on the learning
- Simple simulations
- High effective output
- Cost effective
  - Value for money

AEROTEAM

# PRACTICAL ADVANTAGE

- Less preparation
  - Plug and play
- High activity level
  - People are an active part of their own learning
- High learning effect
  - Fun and educational
- Multiple inputs and solutions
  - Sharing knowledge
- Assessment of performance in real time
  - Comparison to other training

AEROTEAM

# SIMULATION

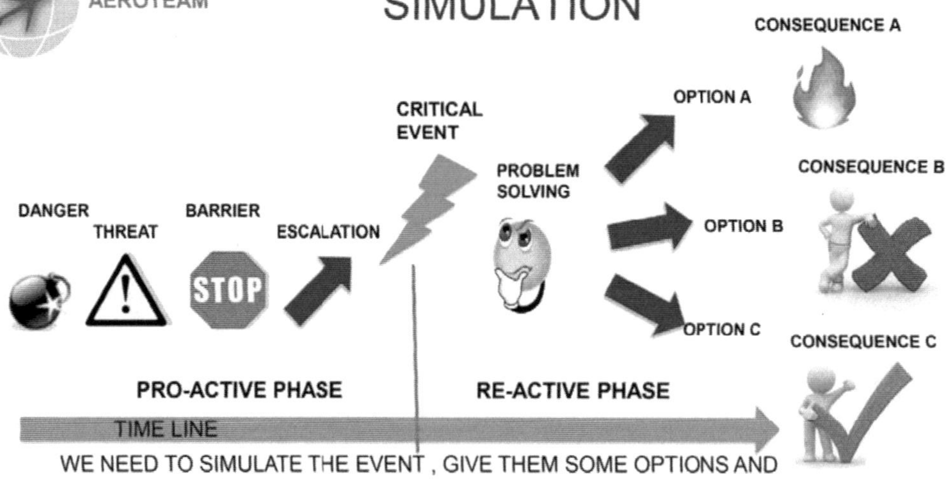


WE NEED TO SIMULATE THE EVENT , GIVE THEM SOME OPTIONS AND THEY NEED TO FIND THE CONSEQUENCE AND SELECT THE ONE THAT IS MOST APPEALING TO THEM.

AEROTEAM

# NOTECHS Categories

**SOCIAL SKILLS**

TEAMWORK

LEADERSHIP & MANAGEMENT SKILLS

**COGNITIVE SKILLS**

SITUATION AWARENESS

DECISION MAKING





AEROTEAM

# TRAINING of NOTECHS

## Assessment and Feedback of Performance

- Operational environment
  - ➢Line training (Normal operation)
  - ➢Simulators (Normal operation / Emergency)
  - ➢Drill Mock-ups (Emergency)
- Non-operational environment
  - ➢Classroom → **Behaviours** – Knowledge – Skills
  - ➢On-line → Knowledge – (Skills)

AEROTEAM

## POSITIVE behaviour

- ✓ Assertive and professional
- ✓ Recognizes signs of stress, fatigue and overload
- ✓ They are open-minded and listen to others
- ✓ Discusses crew capabilities / competence
- ✓ Relaxed, supportive and approachable

## NEGATIVE Behaviour

- ✖ Competes with others
- ✖ Does not show initiative
- ✖ Fails to identify potential threats or possible problems
- ✖ Inappropriate use of humour
- ✖ Indifferent towards other crew members



APATS Presentation
AEROTEAM
VER02-170824/SSA

AEROTEAM

## POSITIVE behaviour

- ✓ Ensures SOP compliance
- ✓ Intervenes if deviation to SOP is observed
- ✓ Creates an atmosphere for open communication
- ✓ Keeps calm and makes suggestions
- ✓ Encourages others to speak up

## NEGATIVE Behaviour

- ✖ Sticks to own opinion without considering a compromise
- ✖ Does not advocate own position
- ✖ Ignores suggestions from other crew members
- ✖ Surprised by consequences they should have anticipated
- ✖ Unaware of Time

© Aeroteam 2017

APATS Presentation
AEROTEAM
VER02-170824/SSA

AEROTEAM

## Requirements for Success

- ✓ Realistic training scenarios
- ✓ Interactive and fun
- ✓ Valid constructive assessment of competence
- ✓ Instructors with good training and assessment skills
- ✓ Valid feedback and critique based on observations
- ✓ Suggestions to improve performance

## Blocks to Success

- ✖ Lack of focus on learning
- ✖ Lack of assessment skills
- ✖ Lack of self-criticism (Reflection on own performance)
- ✖ Training scenario too complex
- ✖ Not providing valid feedback on performance





AEROTEAM

## NOTECHs in Practice

- It is behaviours that are observable and not the actual skills!
- Instructor knowledge of the NOTECHs Behaviour is essential!
- Definition of effective Human Factor behaviour is visible to the crew
  - Helps crew to compare own performance
- In Simulations we are using their technical knowledge to enhance their Non-Technical Skills.
- We focus on "**How they do**" rather than "**What they do**"

AEROTEAM

## OBSERVATIONS

- Crew behaviour
- Communication
- Listening skills
- Do they disagree?
- How do they solve that?
- Is there a "leader"?
- Is someone very passive?
- How do they handle conflicts?
- Do they follow SOP?
- Can they foresee the consequences?
- Does everyone give inputs?
- How do they reach an agreement?
- Is someone dominant / controlling?



APATS Presentation
AEROTEAM
VER02-170824/SSA

AEROTEAM

## LIMITATION of Assessment

- Not every aspect of performance and behaviour can be captured:
  - Limited occurrence of some behaviours
    - i.e. Conflict resolution
  - Limitation of human observers
    - Distraction
    - Overload
    - Large groups
    - Lack of knowledge of NOTECHS
    - Own perception of Level of Performance (Standardization)

© Aeroteam 2017

APATS Presentation
AEROTEAM
VER02-170824/SSA

AEROTEAM

## SUMMARY

- Introduction of Learning Games into crew training will:
  - Enhance learning effect by active participation
  - Enhance the Non-technical Skills of the crew
  - Enhance the knowledge of the operation at many levels
  - Define problem areas for future training (EBT)
  - Enhance instructor's competence in Assessment



APATS Presentation
AEROTEAM
VER02-170824/SSA

AEROTEAM

## SUMMARY

- Introduction of Learning Games into crew training will:
  - Connect classroom training to real world operation
  - Increase the value of the training
  - Provide cost effective classroom training
  - Engage and motivate the participants
  - Provide possibility for Assessment of Performance in the classroom



APATS Presentation
AEROTEAM
VER02-170824/SSA

AEROTEAM

## Summary

- Learning games have to be:

  **Realistic** – a simulation of reality

  **Simple** – simple presentation of complex problems

  **Precise** – clear and precise learning objective

  **Well designed**

  **Measureable** for the learning effect.





AEROTEAM

## THANK YOU FOR YOUR ATTENTION!

Aeroteam
Rørkærvej 108,
8722 Hedensted
+45 20 32 36 76
ssa@aeroteam.dk
www.aeroteam.dk

...if everything seems under control, you're not going fast enough...

Customized Training with Computerized Adaptive Testing
Jian Gu, YanJin Sun ( Eastern Air)
Jasper Li, Yuping Guan (Beijing Pelesys),
Aug 2017
中國東方航空飛行安全技術應用研究院
CHINA EASTERN Flight Safety Technology Institute

About Us
China Eastern Airlines
600+
100 M+
7th worldwide
中國東方航空
CHINA EASTERN
中國東方航空飛行安全技術應用研究院
CHINA EASTERN Flight Safety Technology Institute

## CES Flight Safety & Technology Institute

Founded in 2014, the only research institute in Chinese Airlines

Research
- Frontier tech. of civil aviation
- New training method

Increase
- Training efficiency

Improve
- Safety of flights

PELESYS

中國東方航空飛行安全技術應用研究院
CHINA EASTERN Flight Safety Technology Institute

## Customized Training

### Customized Training Method

Training Records + Human Factors Analysis + QAR Data Analysis = Customized Training (T2P)

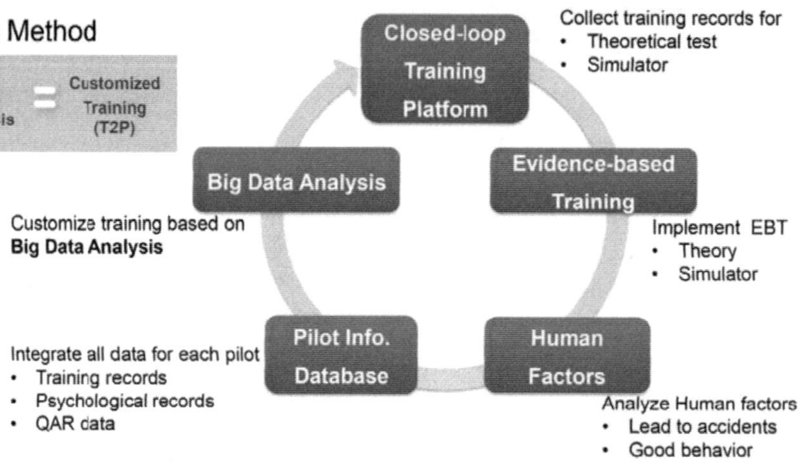


PELESYS

中國東方航空飛行安全技術應用研究院
CHINA EASTERN Flight Safety Technology Institute

## Closed-loop Training Platform

Closed-loop Training Platform Base On Adaptive Exam System

Road Map

**STEP 1**
- A320
- Pilot

→

**STEP 2**
- Establish Adaptive Exam System
  - Individualized Exam
  - CAT
  - Weakness Improvement

→

**STEP 3**
- Expand To A330 A350 B737 B777 B787

→

**STEP 4**
- Expand To Engineers &Flight Attendants

PELESYS

中國東方航空飛行安全技術應用研究院
CHINA EASTERN Flight Safety Technology Institute

## Adaptive Exam System

**Adaptive Exam System is based on The CAT (Computerized Adaptive Testing) Theory**

A. Question Difficulty Analysis
- Question difficulty analysis will be done by the system based on the Rasch model theory, instead of subjective assignment by instructors.

B. Pilot Ability Analysis
- In-depth analysis will done on pilot ability, rather by measuring the raw scores.

C. Exam Quality Analysis
- Each exam will be analyzed to meet quality standards before publishing.

D. Question Quality Analysis
- Each question will be analyzed to ensure the reliablity and effectiveness.

PELESYS

中國東方航空飛行安全技術應用研究院
CHINA EASTERN Flight Safety Technology Institute

## Question Difficulty Analysis

- Rasch model is a psychometric model for analyzing categorical data, such as answers to questions on assessment or questionnaire response
- The main outcomes are the estimate of students' abilities and question difficulties.

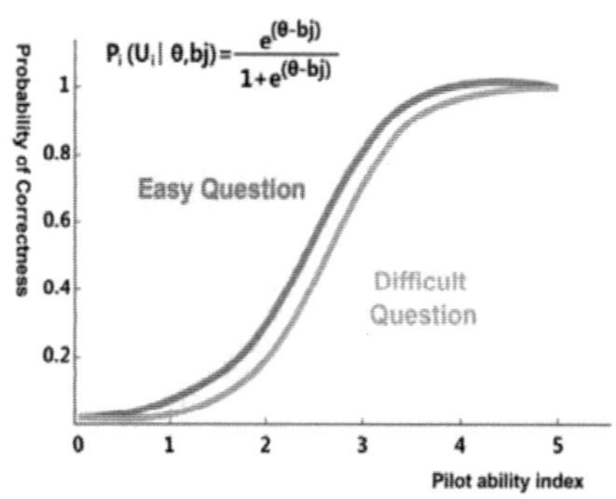


PELESYS

中國東方航空飛行安全技術應用研究院
CHINA EASTERN Flight Safety Technology Institute

## Pilot Ability Analysis

- The following formula is to estimate students' abilities.
- The question difficulty and ability estimates are mutually dependent.

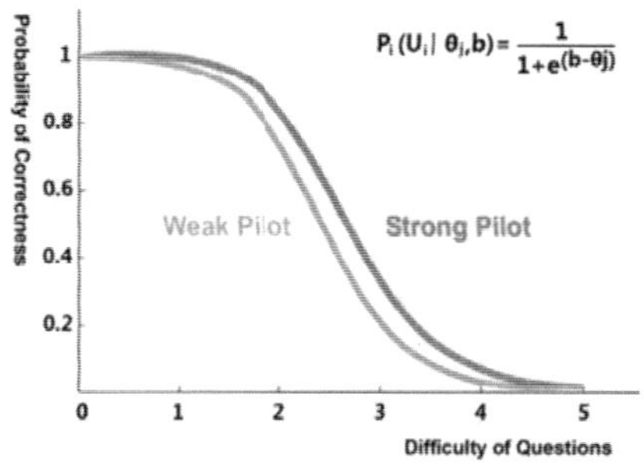


PELESYS

中國東方航空飛行安全技術應用研究院
CHINA EASTERN Flight Safety Technology Institute

## Exam Quality Analysis

- KR20 algorithm is used to measure the quality of an exam.
- If all students get 100% or 0%, KR20 = 0
- If KR20 > 0.7, it means the exam is of good quality.

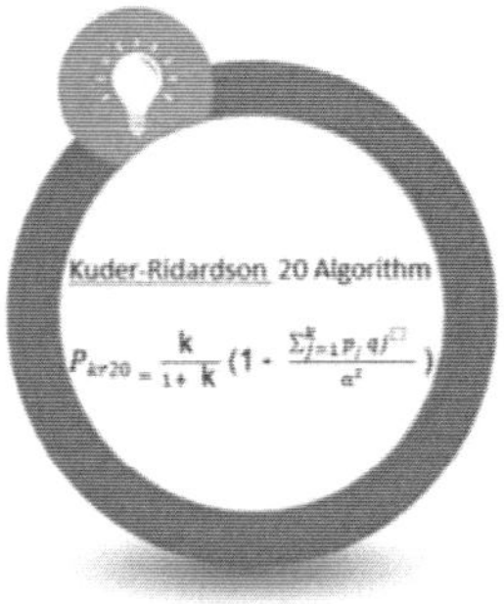


PELESYS

中國東方航空飛行安全技術應用研究院
CHINA EASTERN Flight Safety Technology Institute

## Question Efficiency Analysis

- The main algorithm to measure the question quality is R: Question Discrimination Index.
- If R > 0.25, the question is of good quality.
- There are other statistics analysis such as mean, standard deviation, ANOVA analysis.

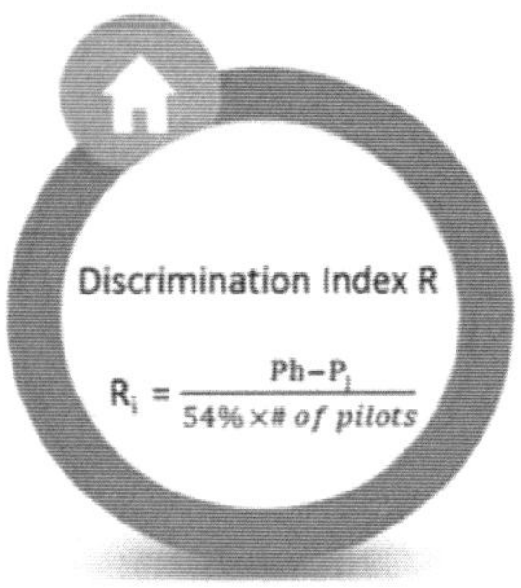



PELESYS

中國東方航空飛行安全技術應用研究院
CHINA EASTERN Flight Safety Technology Institute

## Standardized Pilot Ability & Question Difficulty

- We will use 1 to 5 as standard scores.
- This has the same score structure as grading score of our "Data Based Simulator Assessment System"
- Our ultimate goal is to standardize all other pilot related performances including QAR as Pilot Performance Index (PPI).

PELESYS

中國東方航空飛行安全技術應用研究院
CHINA EASTERN Flight Safety Technology Institute

## Current Research

- We performed an analysis on a previous exam data and the results are as follows.
- KR20 = 0.75, which means this exam is reliable.
- Most R are greater than 0.4, which means the questions are reliable.

### Rasch Analysis

Student abilities and question difficulties are estimated simultaneously.

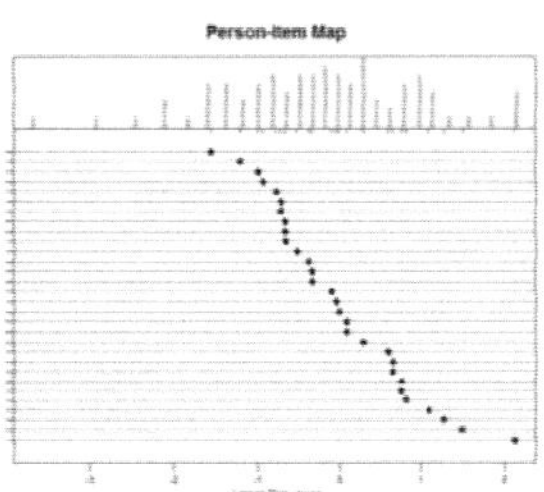


### CAT Testing

The following simulation indicates that CAT only took 40 questions to complete the testing.

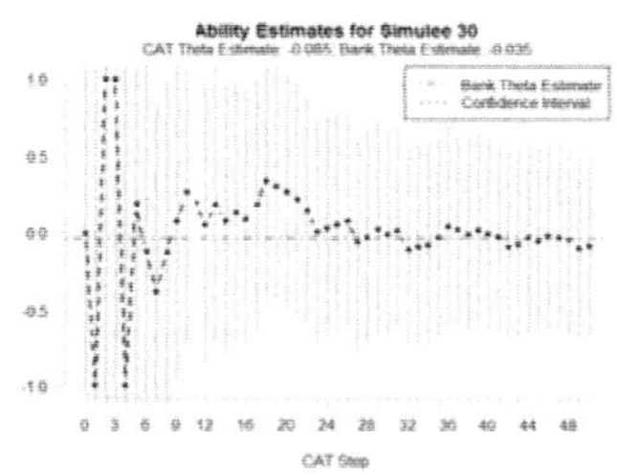


PELESYS

中國東方航空飛行安全技術應用研究院
CHINA EASTERN Flight Safety Technology Institute

中國東方航空飛行安全技術應用研究院
CHINA EASTERN Flight Safety Technology Institute
Thank You !
PELESYS

# JOINING THE DOTS TO IMPROVING PILOT WELLBEING & COGNITIVE FUNCTION

AVIATION COACHING ALLIANCE

ASIA PACIFIC AVIATION TRAINING SYMPOSIUM 2017 SINGAPORE

Margie Burns MBS B.Bus CIPD BPS EAAP

Aviation Selection Consultants


AVIATION SELECTION CONSULTANTS



## Contents

- Stress & stressors affecting pilots - impacting decision-making & well-being.
- Improving cognitive function.
- Elements of a Wellbeing Programme.
- Case Study - pilot wellbeing programme scientifically proven to improve cognitive function.
- Benefits of a Wellbeing Programme.
- Looking Ahead - Where to next?

AVIATION COACHING ALLIANCE

© Copyright of Aviation Selection Consultants

## Stress & Stressors

- Stress is the overwhelming feeling of being under so much pressure that you feel unable to cope.
- It can affect how you think, feel, behave & how your body works.
- It can present itself through a variety of symptoms
  - Mental
  - Physical
  - Behavioural
  - Emotional
- A study conducted by the International Stress Management Association
  - > 50% of working people suffered stress over 1 year
  - 25% had taken time off because of it.
- Given that being a pilot is regarded as one of the most stressful jobs in the world what are we doing or can we do beyond an annual pilot medical check to improve the wellbeing & cognitive function of our pilots.

N W E S NAVIGATE WELLBEING EMPOWER SUCCEED AVIATION COACHING ALLIANCE

## Reality of Modern Day Stressors

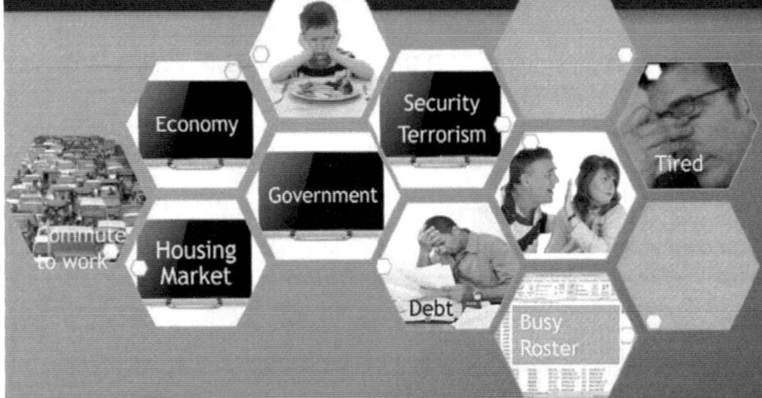

# Stress Responsibility

*'The truth is that stress doesn't come from your boss, your kids, your spouse, traffic jams, health challenges, or other circumstances. It comes from your thoughts about these circumstances'*

Author & Philosopher, Andrew Bernstein

AVIATION COACHING ALLIANCE

# SOURCES OF STRESS FOR PILOTS

- **Physical Stressors**
  - Noise, vibration, humidity, lack of oxygen in cockpit.
- **Physiological Stressors**
  - Fatigue, hunger, dehydration, jet lag, sleeping difficulties, various illnesses.
- **Psychological Stressors**
  - Emotional factors - financial problems, spouse, illness/death of family member, burnout.

# Effects of Stress on Pilot Performance

- **'Pilots are expected to have** the ability to perceive, think & act to the best of their ability without the hindering effects of anger, worry, anxiety.
- Studies have shown that emotional factors, mental upsets & psychological mal-adjustments are repeatedly present in airplane accidents.
- The ability to think clearly & act decisively are greatly influenced by feelings & emotions. Good judgement & **cognitive function** is seriously impaired under stress, fatigue'.
- Source: 'Human Factors in Aviation: The effects of Stress on Pilot Performance'.

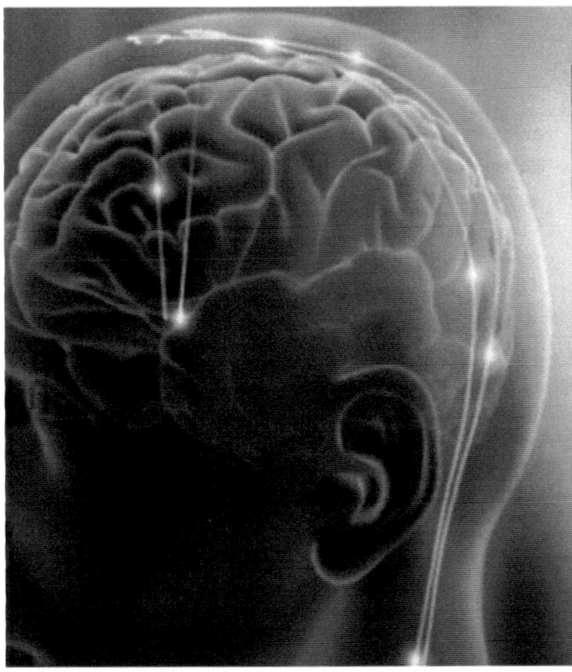

## BRAIN RESEARCH FINDINGS

AFFECTING WELLBEING/COGNITIVE FUNCTION

- Up to 27 years ago brain researchers thought that after puberty that the only changes that take place in the brain are negative - loss of brain cells & brain damage due to stroke etc.
- Neuroplasticity of the brain?
- Every time you learn a new skill or fact, you change your brain.
- Our surroundings & personal experiences don't just affect our beliefs and perceptions. They can physically change our brain.
- While the plasticity of the brain decreases with age, studies have shown remarkable amounts of re-organization in the adult brain & further research shows that our behaviours change our brain & these changes are not limited by age.

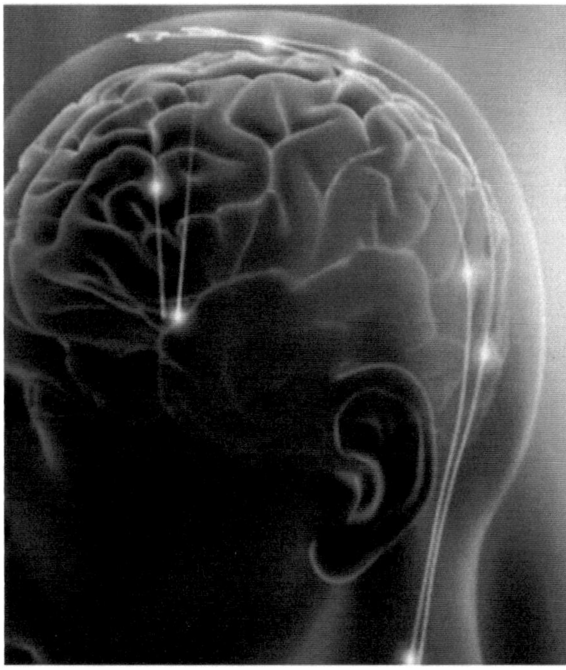

## BRAIN RESEARCH FINDINGS

### AFFECTING WELLBEING/COGNITIVE FUNCTION

- Meditation has been scientifically proven to change the shape of our brain.
- Study Of London Taxi Drivers - revealed they have a larger hippocampus than other humans because they have to memorise street maps.
- A study at University of Kansas revealed you can improve cognitive function in adults over 65 with exercise. Significantly improved - spatial processing & ability to focus - key for pilots.
- The more you stimulate neurons the stronger they get.

## Elements of a Wellbeing Programme

AVIATION COACHING ALLIANCE

- Vitamin Testing & Nutritional Plan
- Exercise Programme
- Sleep Aids/Techniques
- Breathing Techniques
- Mindfulness & Meditation
- Stress Management Awareness Online Course
- One to one coaching sessions
- Pre & Post Programme Cognitive testing

# Vitamin Testing

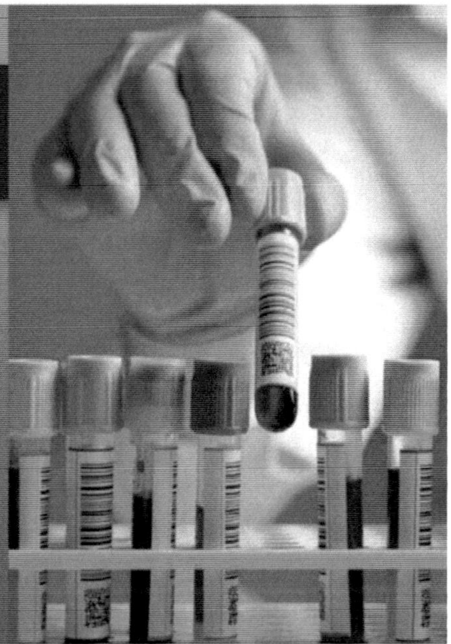

- 17 Essential Vitamins & Minerals for Brain Function
- Main Deficiencies found:
- Magnesium
- Vitamin B Complex
  - Vitamin B12
  - Vitamin B6
- Vitamin C
- Vitamin D
- Vitamin E
- Omega 3

# Nutrition

AVIATION COACHING ALLIANCE

- Many people, when stressed, turn to eating lots of unhealthy food & drink or even stop eating altogether.

A healthy diet should consist of:

- Fruit & Vegetables (plenty of greens)
- Essential Fats such as nuts, seeds, avocados & oily fish
- Water
- Protein, such as fish, eggs, beans & meat.
- Whole grains, such as quinoa & rice (brown preferable)
- Complex carbohydrates such as sweet potato, rye bread.
- Dairy foods - probiotics yogurt, kefir.

## Nutrition

Foods to avoid eating:

- Simple carbohydrates, such as sugar, chocolate, biscuits, cakes.
- High sugar drinks.
- Caffeinated drinks, such as tea, coffee & energy drinks.
- Alcohol.
- Fried food.

## Exercise Programme

- Ask anyone who is stressed whether they want to do some exercise and the answer is likely 'I don't have time' or 'I don't have energy'.
- Even if it's just 20 minutes a day, all exercise has a benefit on your body.
- Exercise can decrease blood pressure and lower heart rate, two of the worst physical symptoms of stress.
- Exercise can also help to change your mood. As you exercise, endorphins are released in your body & help produce feelings of well-being & calmness.
- For increased brain function go for intensity over duration & bump up fitness level.

## Sleep:

- Sleep is one of the most important factors in reducing stress. If your body is tired & stressed, you are likely to become ill, emotional.
- Aiming to get 7-8 hours of a good night's sleep needs to be a priority.
- For cadets with sleep issues - huge success with making personalised MP3 recordings to help them sleep (meditation).
- Providing them with sleep techniques/advice.

## Breathing

- When you are stressed your breathing becomes faster & shallower.
- Not as much oxygen reaches your bloodstream or your brain and so your heart rate & blood pressure increase, you can feel light headed & get a bit shaky.
- Practising relaxation techniques helps regain control and bring breathing back down to its normal, slow and regular state.

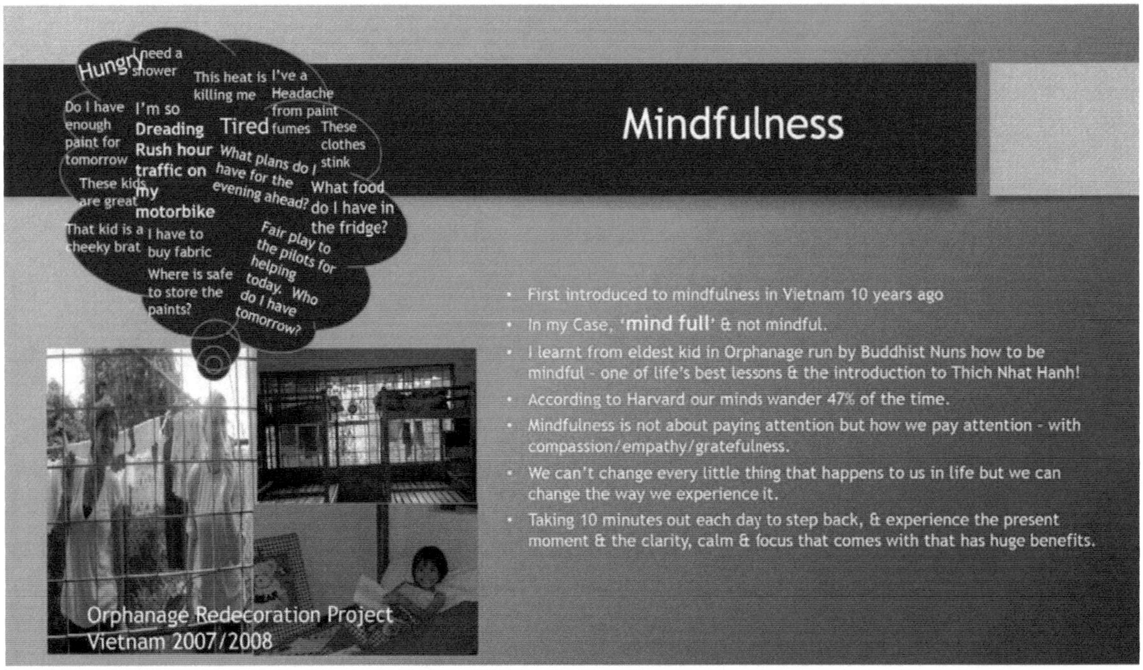
Mindfulness
Hungry
I need a shower
This heat is killing me
I've a Headache from paint fumes
Do I have enough paint for tomorrow
I'm so Dreading Rush hour traffic on my motorbike
Tired
These clothes stink
What plans do I have for the evening ahead?
What food do I have in the fridge?
These kids are great
That kid is a cheeky brat
I have to buy fabric
Fair play to the pilots for helping today. Who do I have tomorrow?
Where is safe to store the paints?
First introduced to mindfulness in Vietnam 10 years ago
In my Case, 'mind full' & not mindful.
I learnt from eldest kid in Orphanage run by Buddhist Nuns how to be mindful - one of life's best lessons & the introduction to Thich Nhat Hanh!
According to Harvard our minds wander 47% of the time.
Mindfulness is not about paying attention but how we pay attention - with compassion/empathy/gratefulness.
We can't change every little thing that happens to us in life but we can change the way we experience it.
Taking 10 minutes out each day to step back, & experience the present moment & the clarity, calm & focus that comes with that has huge benefits.
Orphanage Redecoration Project Vietnam 2007/2008

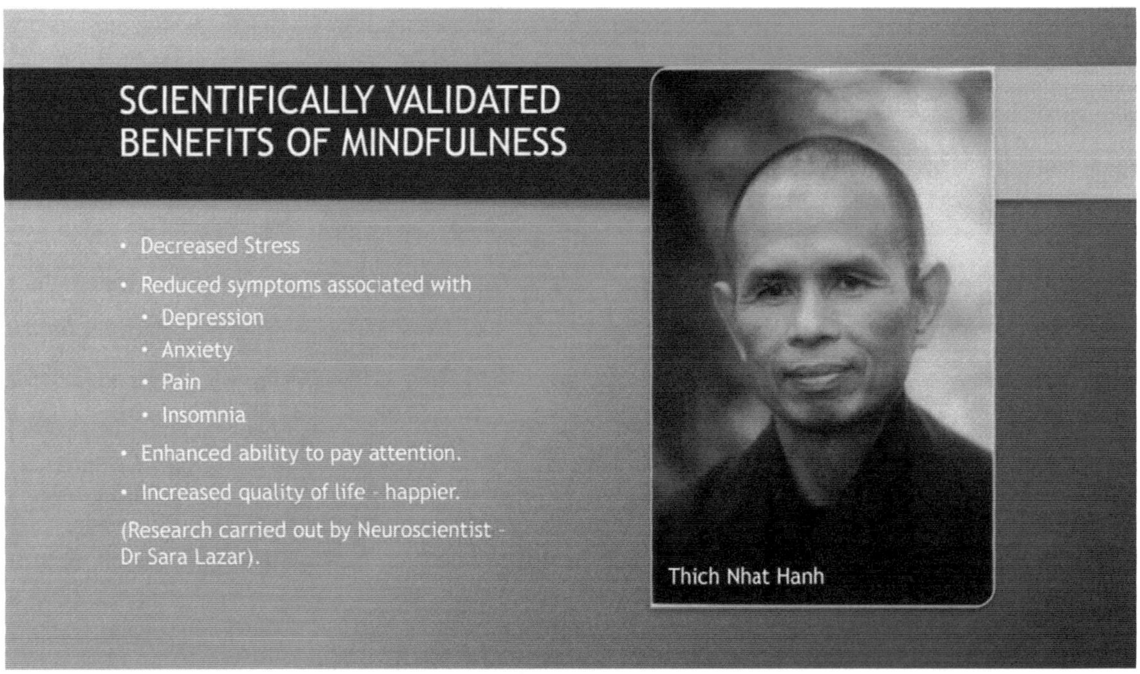
SCIENTIFICALLY VALIDATED BENEFITS OF MINDFULNESS
Decreased Stress
Reduced symptoms associated with
Depression
Anxiety
Pain
Insomnia
Enhanced ability to pay attention.
Increased quality of life - happier.
(Research carried out by Neuroscientist - Dr Sara Lazar).
Thich Nhat Hanh

## RESULTS OF STUDY ON CADET PILOT WELLBEING PROGRAMME

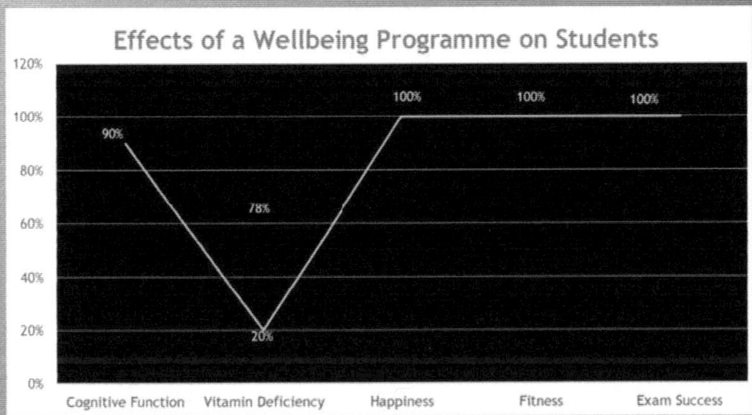


- 90% improved cognitive function
- 10% at 99th Percentile both pre & post.
- 78% of sample had vitamin deficiencies - all with deficiency in (Magnesium & Vitamin B12) & others with Vit D & B6 & Omega 3 deficiencies.
- 100% experienced more happiness
  - Less stress/more positive about future/confidence
- 100% increased fitness level
  - Lower Resting heart rate
    - Ran 1 mile faster
    - 30 Situps
    - 20 Pushups
    - Lower BMI
- 100% passed all exams & attributed success to wellbeing programme.
- 60% Males & 40% Females
- Average age: 26 years
- Sample: 16



## Some Benefits of Having Employee Wellbeing Programme

AVIATION
COACHING
ALLIANCE

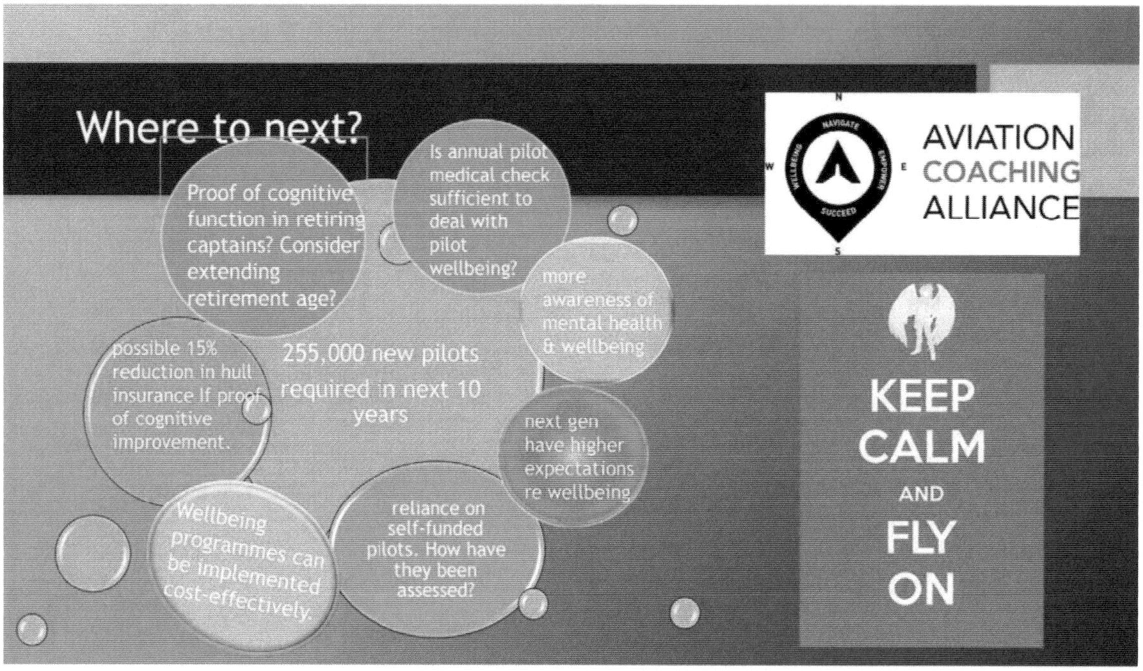
Where to next?
Proof of cognitive function in retiring captains? Consider extending retirement age?
Is annual pilot medical check sufficient to deal with pilot wellbeing?
more awareness of mental health & wellbeing
possible 15% reduction in hull insurance if proof of cognitive improvement.
255,000 new pilots required in next 10 years
next gen have higher expectations re wellbeing
Wellbeing programmes can be implemented cost-effectively.
reliance on self-funded pilots. How have they been assessed?
AVIATION COACHING ALLIANCE
KEEP CALM AND FLY ON

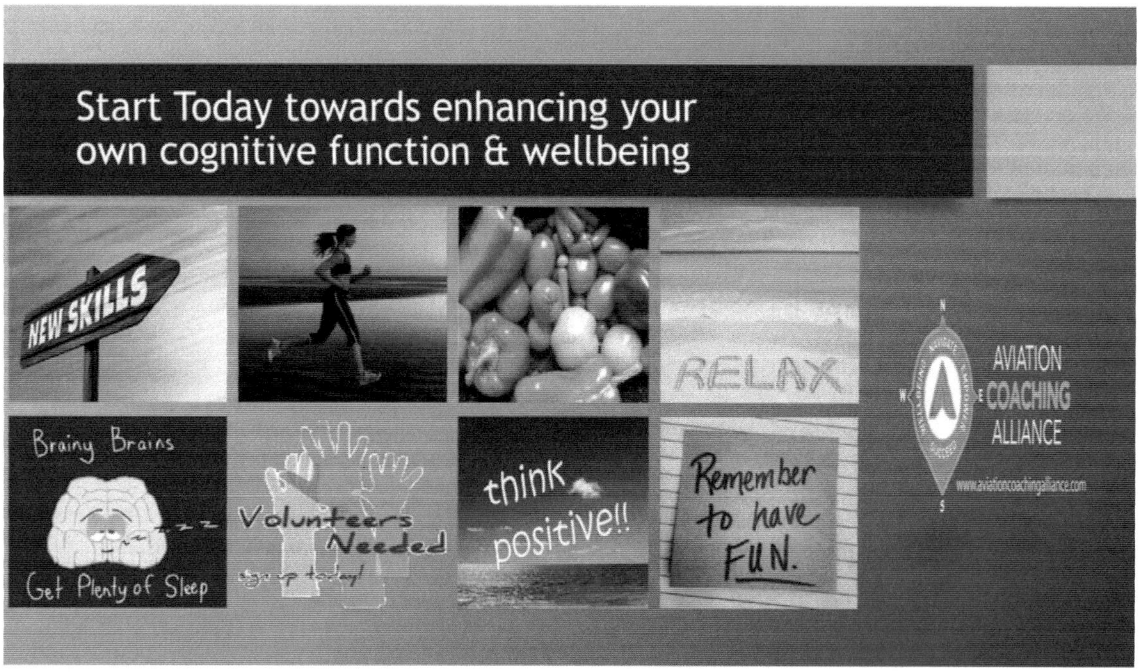
Start Today towards enhancing your own cognitive function & wellbeing
NEW SKILLS
RELAX
Brainy Brains
Get Plenty of Sleep
Volunteers Needed
think positive!!
Remember to have FUN.
AVIATION COACHING ALLIANCE
www.aviationcoachingalliance.com

# References


- NHS Report on Wellbeing
- Dr Lara Boyd, Brain Researcher at University of British Columbia, Canada - TEDx talks.
- Dr Max Cynader, Brain Researcher at University of British Columbia, Canada, TEDx talks.
- Human Factors in Aviation - the effects of stress on pilot's performance.
- Fawne Hansen - "The Adrenal Fatigue Solution". (How do airline pilots cope with stress).
- https://www.careercast.com/content/most-stressful-jobs-2012-3-airline-pilot.
- Aviation Selection Consultants 'Stress Management Awareness Course' material.
- Eric D Vidoni et al: Dose-Response of Aerobic Exercise on Cognitive function. A community based pilot randomised controlled trial. PLOS One. 2015; 10 (7)e 0131647DOI: 10.1371/Journal. Pone.0131647


# Thank You


Margie Burns MBS B.Bus CIPD BPS EAAP
Managing Director
Aviation Selection Consultants
Aviation Coaching Alliance

www.aviationselectionconsultants.com
www.aviationcoachingalliance.com


AVIATION COACHING ALLIANCE
www.aviationcoachingalliance.com

# English language proficiency for successful aviation training

Latitude Aviation English Services

*Henry Emery, Managing director, Latitude Aviation English Services (UK)*

# A

aileron
angle of attack
ATIS
area navigation
azimuth
active runway
APU
anti-skid
aerofoil
airframe
alternate
axis
airspeed
ASI
altitude
approach
ACARS
approach plate
anti icing
attitude
autopilot
AVGAS
ADF
ATC
artificial horizon
avionics

Latitude Aviation English Services

*English language proficiency for successful aviation training, APATS 2017*

## Science Technology Engineering Maths

AVIATION CONTEXT

FOUNDATION LANGUAGE

Latitude Aviation English Services *English language proficiency for successful aviation training, APATS 2017*

**CRIDDLE'S REACTION**

Within the criddle chamber, there are nine graysons. The graysons are spaced at a distance of 5 mm to allow a steady flow of fenaldioxide to pass between them. At the beginning of each cycle, the graysons are charged with between 23,000 and 25,000 c/ps. The super cooled fenaldioxide is then pumped into the chamber. The surface temperature of the graysons increases as a result of the charge. This warms the fenaldioxide which begins to release iridate molecules at 180°C. At approximately 270°C, the iridate molecules begin to vaporise, causing a rapid rise in pressure within the criddle chamber which creates the necessary conditions for Criddle's reaction to take place.

The graysons ...

a. Stop the flow of fenaldioxide

b. Are charged with more than 25,000 c/ps

c. Transfer heat to the fenaldioxide ✓

d. Rise in the Criddle chamb

Latitude Aviation English Services *English language proficiency for successful aviation training, APATS 2017*

## CRIDDLE'S REACTION

Within the criddle chamber, there are nine graysons. The graysons are spaced at a distance of 5 mm to allow a steady flow of fenaldioxide to pass between them. At the beginning of each cycle, the graysons are charged with between 23,000 and 25,000 c/ps. The super cooled fenaldioxide is then pumped into the chamber. The surface temperature of the graysons increases as a result of the charge. This warms the fenaldioxide which begins to release iridate molecules at 180°C. At approximately 270°C, the iridate molecules begin to vaporise, causing a rapid rise in pressure within the criddle chamber which creates the necessary conditions for Criddle's reaction to take place.

Latitude Aviation English Services

*English language proficiency for successful aviation training, APATS 2017*


## CRIDDLE'S REACTION

Within the criddle chamber, there are nine graysons. The graysons are spaced at a distance of 5 mm to allow a steady flow of fenaldioxide to pass between them. At the beginning of each cycle, the graysons are charged with between 23,000 and 25,000 c/ps. The super cooled fenaldioxide is then pumped into the chamber. The surface temperature of the graysons increases as a result of the charge. This warms the fenaldioxide which begins to release iridate molecules at 180°C. At approximately 270°C, the iridate molecules begin to vaporise, causing a rapid rise in pressure within the criddle chamber which creates the necessary conditions for Criddle's reaction to take place.


*pressure — pressurise — pressurisation*


Latitude Aviation English Services

*English language proficiency for successful aviation training, APATS 2017*

sky SPORTS
Get Sky Sports
Sign In
Search Sky Sports
Home
Sports
Score Centre
Video
Sky Bet
More
Watch Sky Sports
Cricket
ICC Cricket World Cup 2015 News
News
Fixtures & Results
Tables
Features
Opinion
Video
Teams
Live On Sky
Betting
Mark Butcher
Cricket Columnist
England v Scotland: World Cup pressure firmly on Eoin Morgan's side
'These are nervous times for out-of-form English players...'
Last Updated: 17/02/15 11:57am
Related Stories
World Cup reborn
Willis backs ICC World Cup plan
Latitude
Aviation English Services
English language proficiency for successful aviation training, APATS 2017

FORTUNE
NBC's Rachel Dolezal interview shows what's wrong with the TV business
NEWS
POPULAR
VIDEOS
FORTUNE 500
TECH
AMAZON
Amazon to start paying more EU taxes after pressure from regulators
by Mathew Ingram
MAY 26, 2015, 11:11 AM EDT
amazon
Latitude
Aviation English Services
English language proficiency for successful aviation training, APATS 2017

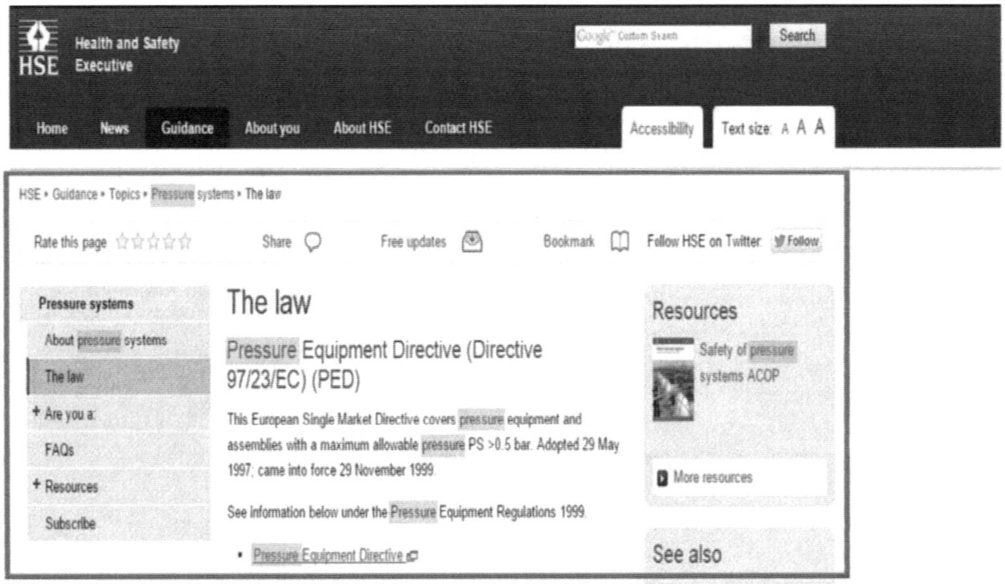

## Principles of flight:

*Because the airflow velocity over the top surface of a lifting aerofoil will be greater than that beneath, the **pressure** differential that results will produce a force per unit area acting upwards*

## Meteorology:

*Expressed simply, atmospheric **pressure**, which acts on any object immersed in the atmosphere, is caused by the weight of air above that object*

Latitude
Aviation English Services

*English language proficiency for successful aviation training, APATS 2017*

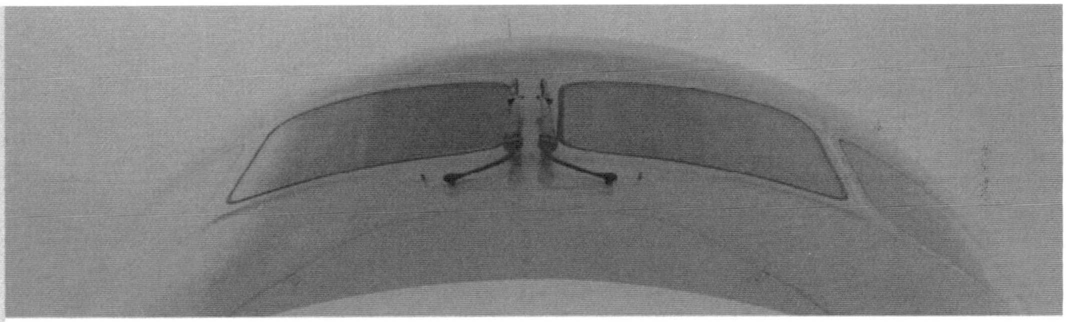

## Airframes:

*The flight deck windows fitted to **pressurised** aircraft must withstand both the loads of **pressurisation** and the impact from birdstrikes*

Latitude
Aviation English Services

*English language proficiency for successful aviation training, APATS 2017*

## Powerplant:

*Because of the divergent shape of the vanes, the* ***pressure*** *of the air increases as it flows outwards*

Latitude Aviation English Services *English language proficiency for successful aviation training, APATS 2017*

## Instrumentation:

*In principle, the simple ASI can be considered as an airtight box divided by a flexible diaphragm, with pitot* ***pressure*** *fed to one side and static* ***pressure*** *to the other side*

Latitude Aviation English Services *English language proficiency for successful aviation training, APATS 2017*

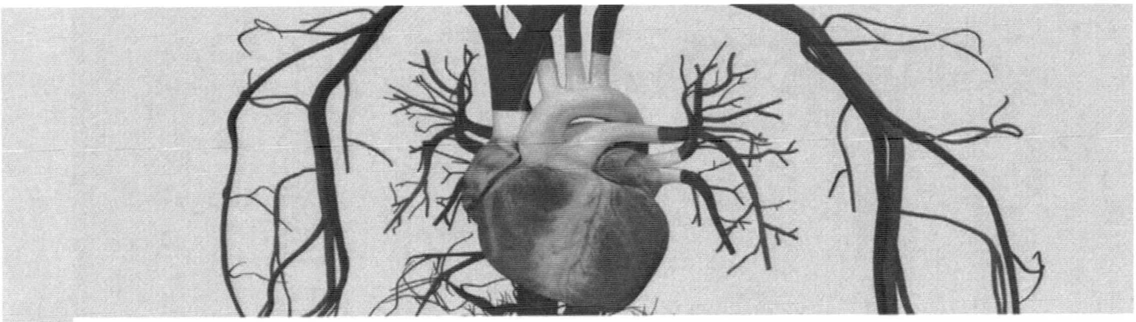

Human performance:

*Arteries carry blood from the heart at high* ***pressure*** *and veins return the blood to the heart at low* ***pressure***

Latitude Aviation English Services *English language proficiency for successful aviation training, APATS 2017*

□□

ciśnienie

πίεση

**pressure**

presión

الضغط

ความดัน

Principles of flight

Meteorology

Airframes

Powerplant

Instrumentation

Human performance

Latitude Aviation English Services *English language proficiency for successful aviation training, APATS 2017*

## FIXED LANDING GEAR

**Rubber Cord.** When rubber cord is used as a shock absorber, the undercarriage is usually in the form of tubular struts, designed and installed so that the landing force is directed against a number of turns of rubber in the form of a grommet or loop.

Oxford Aviation Academy ATPL Ground Training Series, Book 2, Airframes, Systems, Chapter 3, pp 81

Latitude Aviation English Services *English language proficiency for successful aviation training, APATS 2017*

## FIXED LANDING GEAR

**Rubber Cord.** When rubber cord is used as a shock absorber, the undercarriage is usually in the form of tubular struts, designed and installed so that the landing force is directed against a number of turns of rubber in the form of a grommet or loop.

- *Dependent relationship*
- *What you can expect / what normally happens*
- *What something is made up of / how it is constructed*

Latitude Aviation English Services *English language proficiency for successful aviation training, APATS 2017*

## FIXED LANDING GEAR

**Rubber Cord.** When rubber cord is used as a shock absorber, the undercarriage is usually in the form of tubular struts, designed and installed so that the landing force is directed against a number of turns of rubber in the form of a grommet or loop.

- ✓ *Reason*

Latitude Aviation English Services *English language proficiency for successful aviation training, APATS 2017*

## FIXED LANDING GEAR

**Rubber Cord.** When rubber cord is used as a shock absorber, the undercarriage is usually in the form of tubular struts, designed and installed so that the landing force is directed against a number of turns of rubber in the form of a grommet or loop.

Latitude Aviation English Services *English language proficiency for successful aviation training, APATS 2017*

## Zero conditional

| If clause (condition) | main clause (result) |
|---|---|
| When / if + present simple, | present simple |
| *When rubber cord is used …* | *the undercarriage is usually …* |
| *If you apply full flap,* | *drag increases* |
| *If you move the stick to the left,* | *the aircraft rolls to the left* |



## FIXED LANDING GEAR

**Rubber Cord.** When rubber cord is used as a shock absorber, the undercarriage is usually in the form of tubular struts, designed and installed so that the landing force is directed against a number of turns of rubber in the form of a grommet or loop.

Latitude Aviation English Services
*English language proficiency for successful aviation training, APATS 2017*

## The passive

| Object + | 'be' + | past participle |
|---|---|---|
| *rubber cord* | *is* | *used* |
| *landing force* | *is* | *directed* |
| *tubular struts* | *are* | *designed and installed* |



## FIXED LANDING GEAR

**Rubber Cord.** When rubber cord is used as a shock absorber, the undercarriage is usually in the form of tubular struts, designed and installed so that the landing force is directed against a number of turns of rubber in the form of a grommet or loop.

**Reduced relative clauses**

~~that are~~

The undercarriage is in the form of tubular struts, designed and installed so that …

Latitude Aviation English Services *English language proficiency for successful aviation training, APATS 2017*

**FIXED LANDING GEAR**

**Rubber Cord.** When rubber cord is used as a shock absorber, the undercarriage is usually in the form of tubular struts, designed and installed so that the landing force is directed against a number of turns of rubber in the form of a grommet or loop.

- ✓ *1 occurrence of the zero conditional*
- ✓ *3 occurrences of the passive voice*
- ✓ *1 reduced relative clause*

Latitude Aviation English Services *English language proficiency for successful aviation training, APATS 2017*

*How clear is the instructor?*

Latitude
Aviation English Services

*English language proficiency for successful aviation training, APATS 2017*

*OK … uhh … let's see … where do we go from here? ... uhm … mmm ... yeah probably best …*

Latitude
Aviation English Services

*English language proficiency for successful aviation training, APATS 2017*

*Let's uh start by going over what we were talking about last week.*

*Does anyone remember what we talked about last Friday?*

Latitude
Aviation English Services

*English language proficiency for successful aviation training, APATS 2017*

*Is there anybody out there?*

- *Is there anybody in this room?*
- *Can anyone answer my question?*
- *You're all on a different planet today!*

Latitude
Aviation English Services

*English language proficiency for successful aviation training, APATS 2017*

*You may recall the Coriolis effect?*

*This is the force equal to … ?*

Do you remember the Coriolis effect?

What is this force is equal to?

Latitude
Aviation English Services

*English language proficiency for successful aviation training, APATS 2017*

*Let's start from scratch with a picture on the board.*

Latitude
Aviation English Services

*English language proficiency for successful aviation training, APATS 2017*

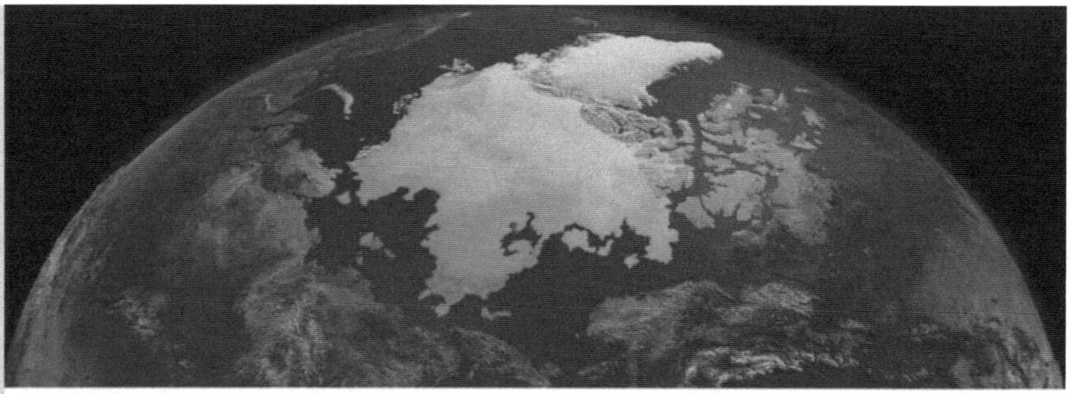

*The North Pole is rotating counter clockwise ... in a direction opposite to a clock, like this ...*

Latitude
Aviation English Services

*English language proficiency for successful aviation training, APATS 2017*

| Verbs | *cause, spin, rotate, stand, stay, view* |
|---|---|

Latitude
Aviation English Services

*English language proficiency for successful aviation training, APATS 2017*

*The Coriolis effect is caused by the spinning of the earth*

*The spinning of the earth causes the Coriolis effect*

Latitude Aviation English Services *English language proficiency for successful aviation training, APATS 2017*

*If you ... were standing on the equator with the sun directly above you, it would be? What time would it be?*

Latitude Aviation English Services *English language proficiency for successful aviation training, APATS 2017*

## Second conditional (unreal hypotheses)

| If clause (condition) | main clause (result) |
|---|---|
| If + past simple, | would + infinitive |
| *If you were standing …* | *what time would it be?* |
| *If we always had direct routings,* | *we would save much more fuel* |
| *If there were a third runway at LHR,* | *approaches would be much quicker* |

Latitude Aviation English Services

*English language proficiency for successful aviation training, APATS 2017*

Delivery (false starts, hesitations, repetitions)

Grading, directness

Organising speech: ► signposts

► voice and pausing

Using paraphrase and gestures

Rate of speech

**FOUNDATION LANGUAGE**

Familiarity with: ► subject matter

► instructor style

► accent

► culture

Latitude Aviation English Services

*English language proficiency for successful aviation training, APATS 2017*

A

aileron
angle of attack
ATIS
area navigation
azimuth
active runway
APU
anti-skid
aerofoil
airframe
alternate
axis
airspeed
ASI
altitude
approach
ACARS
approach plate
anti icing
attitude
autopilot
AVGAS
ADF
ATC
artificial horizon
avionics



Science Technology Engineering Maths

AVIATION CONTEXT

FOUNDATION LANGUAGE

Activity 1: Read the text about Criddle's reaction and answer the question below.

**CRIDDLE'S REACTION**

Within the Criddle chamber, there are nine graysons. The graysons are spaced at a distance of 5 mm to allow a steady flow of fenaldioxide to pass between them. At the beginning of each cycle, the graysons are charged with between 23,000 and 25,000 c/ps. The super cooled fenaldioxide is then pumped into the chamber. The surface temperature of the graysons increases as a result of the charge. This warms the fenaldioxide which begins to release iridate molecules at 180°C. At approximately 270°C, the iridate molecules begin to vaporise, causing a rapid rise in pressure within the Criddle chamber, thus creating the necessary conditions for Criddle's reaction to take place.

The graysons ...

a. Stop the flow of fenaldioxide

b. Are charged with more than 25,000 c/ps

c. Transfer heat to the fenaldioxide

d. Rise in the Criddle chamber

Activity 2: Read the text about Fixed Landing Gear. Underline the words that are **only** about aviation.

**FIXED LANDING GEAR**

**Rubber Cord.** When rubber cord is used as a shock absorber, the undercarriage is usually in the form of tubular struts, designed and installed so that the landing force is directed against a number of turns of rubber in the form of a grommet or loop.

SOURCE: Oxford Aviation Academy ATPL Ground Training Series, Book 2, Airframes, Systems, Chapter 3, pp 81

Activity 3: Listen to the lesson. Decide how clear the instructor is (1 = not clear at all, 5 = very clear).

1 ☐ 2 ☐ 3 ☐ 4 ☐ 5 ☐

Morning! // OK … uhh … let's see … where do we go from here? ... uhm … mmm ... yeah probably best … let's uh start by going over what we were talking about last week. Does anyone remember what we talked about last Friday? Ladies and gentlemen? Is there anybody out there? A clue then. Something caused by the spinning of the earth. Anyone? You may recall the Coriolis effect? // Right! This is the force equal to … ? OK. This is the force is equal to two times omega rho vee sine theta. Remember … ? // OK. Let's start from scratch with a picture on the board. Here is the earth … // and this is the North Pole // and this … is the South Pole. // Viewed from above, the North Pole … uhh … the North Pole is rotating counter clockwise … in a direction opposite to a clock, like this. // If you … were standing on the equator with the sun directly above you, it would be …? What time would it be? // Excellent! // Now, you stay standing there on the rotating earth … how long would it be …

Thank you!
henry@latitude-aes.aero
Latitude
Aviation English Services
Henry Emery, Managing director, Latitude Aviation English Services (UK)

## FROM CLASSROOM TO FTD
## – THE VERSATILE FLAT PANEL TRAINER

**Mark Dransfield**
FRAeS, CEng
Director Regulatory Affairs

TRU SIMULATION + TRAINING
A Textron Company

---

### FROM CARDBOARD BOMBERS & PAPER TIGERS…

TRU SIMULATION + TRAINING

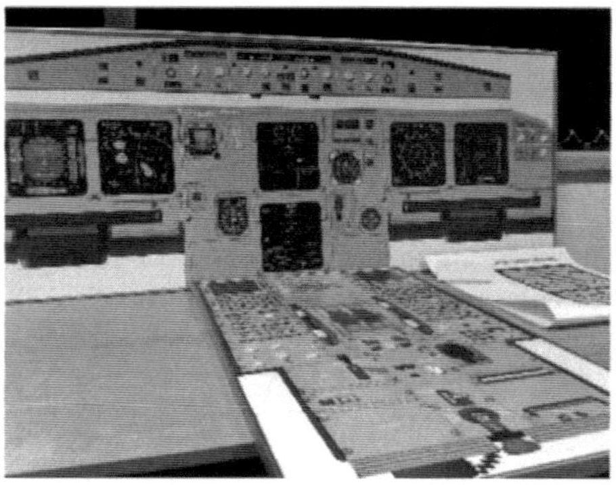

2

## TO TODAYS FAR MORE ADVANCED & CAPABLE DEVICES

TRU SIMULATION + TRAINING

MFTD = **M**aintenance & **F**light **T**raining **D**evice
FPT = **F**lat **P**anel **T**rainer
VPT = **V**irtual **P**anel **T**rainer
APT = **A**dvanced or **A**ircraft **P**rocedures **T**rainer
CPT = **C**ockpit **P**rocedures **T**rainer

FTD = **F**light **T**raining **D**evice
(EASA L1, L2) (FAA Level 1 to 7)

3

## B737 MAX FPT

TRU SIMULATION + TRAINING

4

## B737 MAX CLASSROOM / FPT

TRU SIMULATION + TRAINING

5

## FUTURE OF FPT....?? DISRUPTIVE TECHNOLOGIES

TRU SIMULATION + TRAINING

6

## FPT/FTD FIDELITY 20+ YEARS AGO...

TRU SIMULATION + TRAINING

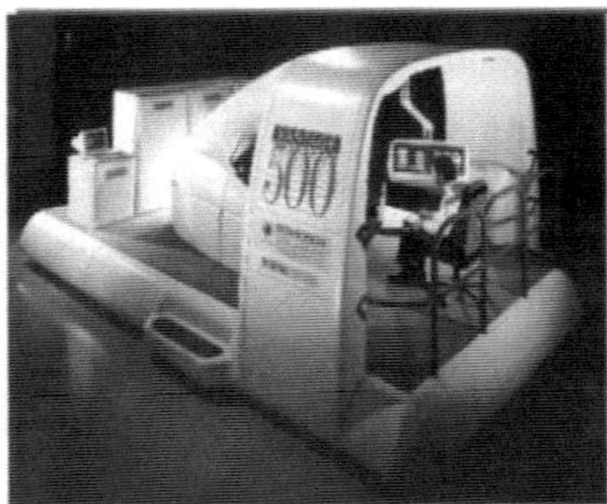

- 1990's
  - Graphics limitations
    - Expensive
  - Bespoke display limitations
    - Poor resolution
  - Usually single touch
    - Reduced capability
  - Reverse engineering
    - a/c systems limited functionality
    - Not truly free play
  - No visual cueing
  - No vestibular cueing
  - Limited Training value
    - CRM, SOPs, Maintenance

7

## FPT/FTD FIDELITY TODAY

TRU SIMULATION + TRAINING

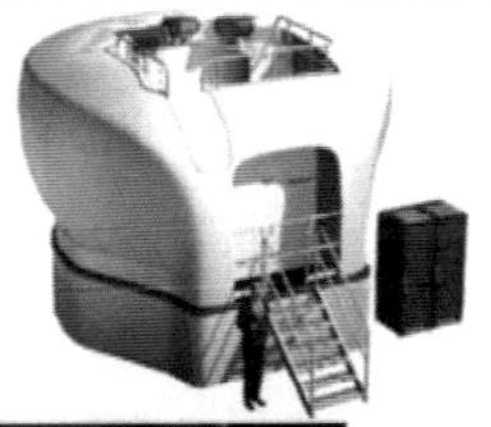

- TODAY
  - High performance graphics cards
    - Driven by gaming industry
  - COTS display solutions
    - Multi touch capability
  - FFS instances with full free play
    - Avionics rehosts
    - HUD / EFB compatibility
  - Visual cueing
    - Direct and collimated displays
  - Vestibular Cueing
    - Motion seats
  - Same fidelity as FFS in terms of:-
    - Aircraft systems modelling
    - Normal and abnormal SOPs
    - Handling and performance
  - Enhanced Training Capability

8

## TRAINING CAPABILITY VS COST (1990'S VS 2017)

TRU SIMULATION + TRAINING A Textron Company

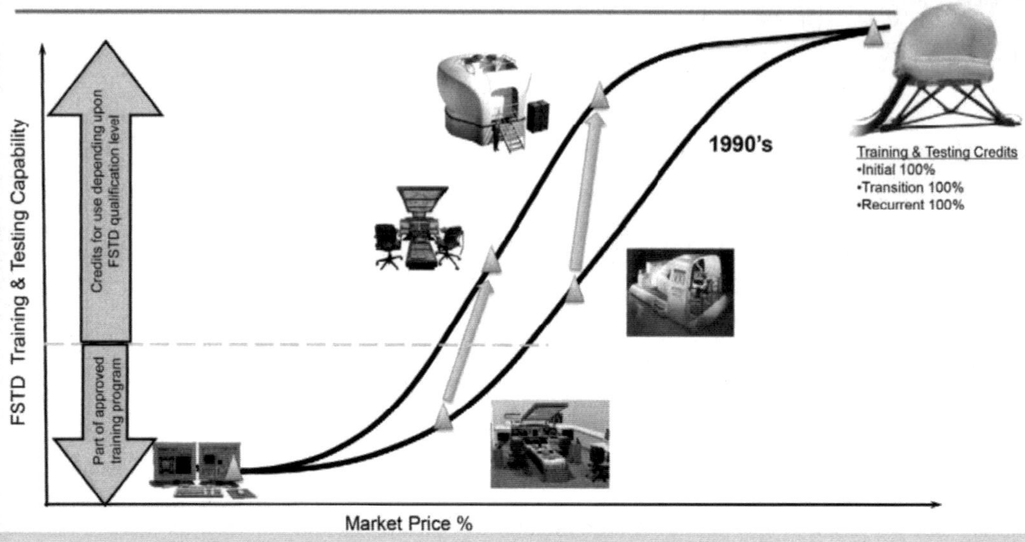


## REGULATORY EVOLUTION (EASA)...FSTD IMPACT

TRU SIMULATION + TRAINING A Textron Company

**EASA NPA 2017-13**

- This NPA proposes to...

  ....introduce guidance on the additional (non-mandatory) capability of each (type of) FSTD to assist stakeholders in defining the use of the appropriate FSTDs to support the training course syllabus as well as the CA in performing the evaluation.

European Aviation Safety Agency

**Notice of Proposed Amendment 2017-13**

Update of flight simulation training devices requirements

**Upset prevention and recovery training, FSTD inspector competencies framework, training matrix**

RMT.0196

**EXECUTIVE SUMMARY**

Rulemaking task (RMT).0196 'Update of flight simulation training devices requirements' has been divided into three work packages (WP1, WP2, and WP3). According to Rulemaking Programme (RMP) 2017-2021, the related documents are planned to be published between 2017 and 2019, in order to complete the modernisation of the certifications specifications (CSs) and acceptable means of compliance/guidance material (AMC/GM) on flight simulation training devices (FSTDs). The objective of this NPA (WP1) is to update CS-FSTD(A), as well as AMC/GM to Annex I (Part-FCL) and Annex VI (Part-ARA) to Regulation (EU) No 1178/2011 (the 'Aircrew Regulation'), to reflect the technological advancements in the field of FSTDs whilst ensuring that FSTD provisions and AMC/GM are fully in line with the current and proposed new training requirements and AMC/GM for flight crew (FC). Furthermore, it aims at supporting the competent authorities (CAs), FSTD operators, and training organisations, by providing a competencies framework for FSTD inspectors as well as guidance for the use of each type of FSTD based on the training need.

This NPA proposes to:

— increase the fidelity of the provisions in order to support the approach-to-stall training, as well as of the new upset prevention and recovery training (UPRT) requirements as proposed by Opinion 06/2017 (RMT.0581);
— increase the fidelity of the simulation of the engine and airframe icing effects;
— develop and deploy an instructor operating station (IOS) feedback tool;
— introduce an FSTD inspector competancies framework; and
— introduce guidance on the additional (non-mandatory) capability of each FSTD to assist stakeholders in defining the use of the appropriate FSTDs to support the training course syllabus as well as the CA in performing the evaluation.

10

## RMT.0196 –UPDATING FSTD REQUIREMENTS
## EASA NOTICE OF PROPOSED AMENDMENT 2017-13
## FSTD CREDITS – AB INITIO TRAINING

**3.2. Draft AMC/GM (draft EASA Decision)**

**3.2.1 AMC/GM to Part-FCL**

1. New GM1 Appendix 9 to Part-FCL is inserted as follows:

**Draft resulting text**

**GM1 Appendix 9 to Part-FCL**

Aeroplane FSTD training credits

| | EASA Training Provisions (hours) | | | | | Maximum FSTD Training Credits (hours) | | | | | | | | | |
|---|---|---|---|---|---|---|---|---|---|---|---|---|---|---|---|
| | AIRCREW/AIROPS reference | Flight instruction [illegible] | Dual | SPIC | Solo (PIC) | BITD | FNPT I | FNPT II | FNPT II MCC | FTD 1 | FTD 2 | FFS Level A/AG | FFS Level B/BG | FFS Level C, CG, interim C | FFS Level D/DG |
| **Type of Training** | | | | | | | | | | | | | | | |
| **LAPL(A)** | | | | | | | | | | | | | | | |
| Modular | FCL.115 | 30 | 15 | NO | 6 | NO | NO | NO | NO | NO | [illegible] | NO | NO | NO | NO |
| **PPL(A)** | | | | | | | | | | | | | | | |
| Modular | FCL.210.A | 45 | 25 | NO | 10 | YES (5) | YES (5) | YES (5) | YES (5) | YES (5) | [illegible] | YES (5) | YES (5) | YES (5) | YES (5) |
| **CPL(A) modular** | | | | | | | | | | | | | | | |
| Instrument flight instruction | Appendix 3, Part E | 10 | 10 | NO | NO | YES (5) | YES (5) | YES (5) | YES (5) | NO | [illegible] | YES (5) | YES (5) | YES (5) | YES (5) |
| **CPL(A) integrated** | | | | | | | | | | | | | | | |
| Instrument flight instruction | Appendix 3, Part D to Annex 1 | 10 | 10 | NO | NO | NO | YES (5) | YES (5) | YES (5) | [illegible] | [illegible] | [illegible] | YES (5) | YES (5) | YES (5) |
| **CPL/IR(A) integrated** | | | | | | | | | | | | | | | |
| Instrument rating training | FCL.315 | 50 | 50 | 20 | NO | NO | YES (25) | Yes (40) | YES (40) | [illegible] | [illegible] | YES (40) | NO | NO | NO |
| **ATPL(A)** | | | | | | | | | | | | | | | |
| Integrated (instrument training) | [illegible] | [illegible] | [illegible] | [illegible] | [illegible] | [illegible] | [illegible] | [illegible] | [illegible] | [illegible] | YES (40) | YES (40) | YES (40) | YES (40) | YES (40) |
| Integrated (MCC training) | [illegible] | [illegible] | [illegible] | [illegible] | [illegible] | [illegible] | [illegible] | [illegible] | [illegible] | [illegible] | NO | YES (15) | YES (15) | YES (15) | YES (15) |
| **MPL integrated** | | | | | | | | | | | | | | | |
| Phase 1 | FCL.410.A + Appendix 5 to Annex 1 | | | | | YES | YES | NO | NO | NO | NO | NO | NO | NO | NO |
| Phase 2 | FCL.410.A + Appendix 5 to Annex 1 | | | | | NO | NO | NO | YES | NO | NO | NO | NO | NO | NO |
| Phase 3 | FCL.410.A + Appendix 5 to Annex 1 | | | | | NO | NO | NO | NO | NO | NO | NO | YES | YES | YES |

11

## RMT.0196 –UPDATING FSTD REQUIREMENTS
## EASA NOTICE OF PROPOSED AMENDMENT 2017-13
## FSTD CREDITS – TYPE RATING, RECENCY, RECURRENT

| | EASA Training Provisions (hours) | | | | | Maximum FSTD Training Credits (hours) | | | | | | | | | |
|---|---|---|---|---|---|---|---|---|---|---|---|---|---|---|---|
| | AIRCREW/AIROPS reference | Flight instruction [illegible] | Dual | SPIC | Solo (PIC) | BITD | FNPT I | FNPT II | FNPT II MCC | FTD 1 | FTD 2 | FFS Level A/AG | FFS Level B/BG | FFS Level C, CG, interim C | FFS Level D/DG |
| **Type of Training** | | | | | | | | | | | | | | | |
| Phase 4 | FCL.410.A + Appendix 5 to Annex1 | | | | | NO | NO | NO | NO | NO | [illegible] | [illegible] | [illegible] | [illegible] | YES |
| **Class Rating(A)** | | | | | | | | | | | | | | | |
| TMG (extension to LAPL(A)) | FCL.725.A | 3 | 3 | NO | NO | NO | NO | NO | NO | NO | [illegible] | [illegible] | [illegible] | [illegible] | NO |
| SEP (Land) | FCL.725.A | | | NO | NO | NO | NO | NO | NO | NO | [illegible] | [illegible] | [illegible] | [illegible] | NO |
| SEP (Sea) | FCL.725.A | 10, 8 if SEP (Land) rated | 10, 8 if SEP (Land) rated | NO | NO | NO | NO | NO | NO | [illegible] | [illegible] | [illegible] | [illegible] | [illegible] | [illegible] |
| MEP (Land) | FCL.725.A | 6 | 6 | NO | NO | NO | NO | NO | NO | NO | [illegible] | [illegible] | [illegible] | [illegible] | [illegible] |
| MEP (Sea) | FCL.725.A | 10, 8 if MEP (Land) rated | 10, 8 if MEP (Land) rated | NO | NO | NO | NO | NO | NO | NO | NO | [illegible] | [illegible] | NO | NO |
| SET | FCL.725.A | | | | | | | | | | | | | | |
| MET | FCL.725.A | | | | | | | | | | | | | | |
| **Type Rating** | | | | | | | | | | | | | | | |
| Type Rating Course SP | [illegible] | [illegible] | [illegible] | [illegible] | [illegible] | [illegible] | [illegible] | [illegible] | [illegible] | YES | YES | YES | YES | YES | YES |
| Type Rating Course MP | [illegible] | [illegible] | [illegible] | [illegible] | [illegible] | [illegible] | [illegible] | [illegible] | [illegible] | PARTIALLY | PARTIALLY | PARTIALLY | YES | YES | YES |
| ZFT Type Rating Course | [illegible] | [illegible] | [illegible] | [illegible] | [illegible] | [illegible] | [illegible] | [illegible] | NO | PARTIALLY | PARTIALLY | NO | NO | YES | YES |
| **Recency of Experience** | | | | | | | | | | | | | | | |
| Recent Experience | FCL.060(b) | NO | NO | NO | NO | NO | NO | NO | NO | NO | NO | NO | YES | YES | YES |
| **Operator Recurrent Training** | | | | | | | | | | | | | | | |
| Route/Area/ Aerodrome Knowledge | AMC1 ORO.FC.105(b)(2);(c)(a),(2)(ii) + (c)(2) | | | | | NO | NO | NO | NO | NO | YES* | YES | YES | YES | YES |
| CRM Training | AMC 1 ORO.FC.115(a)(1) + (a)(4) | | | | | NO | NO | NO | NO | NO | YES* | YES | YES | YES | YES |
| Command Course | ORO.FC.205 | | | | | NO | NO | NO | NO | NO | NO | PARTIALLY | YES | YES | YES |
| Operator Conversion ZFTT (6 circling patterns) | ORO.FC.220(e)(2) | | | | | NO | NO | NO | NO | NO | NO | NO | NO | YES | YES |
| Operator Conversion Training and Checking | AMC 1 OROR.FC.220(a)(iii) + (d)(4) | | | | | NO | NO | NO | NO | NO | PARTIALLY | PARTIALLY | PARTIALLY | YES | YES |

12

RMT.0196 –UPDATING FSTD REQUIREMENTS
EASA NOTICE OF PROPOSED AMENDMENT 2017-13
FSTD CREDITS – INSTRUCTOR TRAINING

BUT…

This guidance for FSTD credits is still based on FTD qualification levels derived from technical standards and training philosophy developed over 20 years ago and not for the capability reflected in todays' modern FTDs/FPTs with…

- Full a/c systems functionality
- FFS performance and handling fidelity
- Collimated visual displays
- Seat based motion cueing

**The deadline for submission of comments on NPA 2017-13 to EASA is 25th September 2017**

13

FSTD ACTUAL TRAINING CAPABILTY (TODAY)

| Training Capability<br>Partially Supports ◔<br>Significantly Supports ◑<br>Fully Supports ● | | FPT  | FTD  | FFS  |
|---|---|---|---|---|
| Airline<br>* Alternate OPC or LST conducted on FFS | EASA FSTD Qual Level | FTD Level I | FTD Level 2 / FNPT II MCC | FFS Level C/D |
| | Transition | ◔ | ● | ● |
| | Recurrent | ◔ | * ● | ● |
| | Recency | | ◑ | ● |
| | Command Upgrade | | ● | ● |
| | Type Rating | ◔ | * ● | ● |
| | ZFT | ◔ | ◑ | ● |

14

## RMT.0599 –EBT –WP02
## COMPETENCY BASED TRAINING REQUIREMENTS FOR ALL LICENCES, RATINGS

### ICAO PAN-TRNG (9868)
#### APPLICABILITY AND AIMS

**2.1 APPLICABILITY**

2.1.1 This manual is intended to provide guidance to civi aviation authorities, operators and approved training organizations in the recurrent assessment and training of pilots conducted in flight simulation training devices (FSTDs), referred to in Annex 6 to the Convention on International Civi Aviation, *Operation of Aircraft*, Part I, *International commercial Air Transport — Aeroplanes*, 9.3, Flight crew member training programmes, and 9.4.4, *Pilot proficiency checks* in addition to Annex 1 to the Convention on International Civil Aviation, *Personnel Licensing*, 1.2.5, Validity of licenses. It may also provide guidance for approved training organizations engaged in the recurrent assessment and training of flight crew operating large or turbojet aeroplanes in accordance with Annex 6, Part II — *International General Aviation — Aeroplanes* (Section 3 refers).

2.1.2 This manual includes guidance for the development of training programmes and the assessment of flight crew performance, in addition to information for instructors conducting the training. The purpose of training and assessment within this programme is to establish competency and provide for continuous and measurable improvement in flight crew performance during line operations.

2.1.3 Aeroplanes considered for application of the guidance in this manual are those with a certified seating capacity of 50 or more passengers for turbo-jet aeroplanes referred to in this manual as *jets*, and of 30 or more passengers for turbo-propeller aeroplanes referred to in this manual as turboprops.

2.1.4 This manual does not formally consider training media, but will assume that the assessment and training described will be conducted in an FSTD qualified to an appropriate level in accordance with CAA rules (Doc 9625 — *Manual of Criteria for the Qualification of Flight Simulation Training Devices* refers). It is recognized that there will be aspects of any recurrent training programme which are conducted outside the environment of a qualified FSTD, but the EBT programme described in this manual is intended to relate only to FSTD assessment and training.

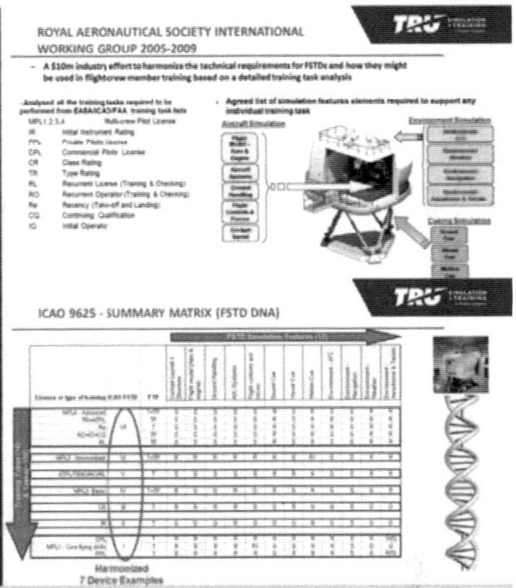


## EBT & COMPETENCIES – FPT/FTD OR FFS?

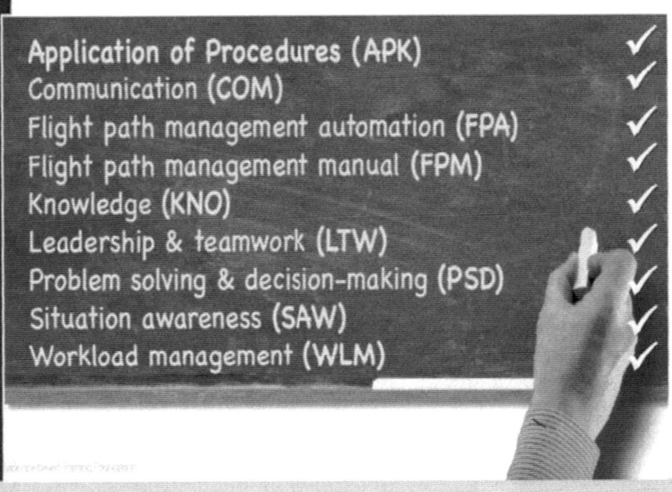


- How many of these competencies can we use CBT/FPT/FTD in addition to a FFS to train or train to proficiency during initial/recurrent training and successfully measure observable behaviours for the desired outcomes?
- Probably a large part of all of them depending on your EBT implementation
- So why is there currently no equivalent FPT/FTD/FFS suitable usage guidance for EBT competencies in FSTD regulatory material?

16

## SUMMARY

- Problems
  - Credit confusion for use of FSTDs below FFS for existing training methodologies and license types
    - still based upon outdated lower level device qualification levels instead of device training capability
  - EBT competencies for training not mapped into any FSTD regulatory guidance that clarify "use of appropriately qualified devices"
- Possible Solutions
  - Credits or training useage guidance based on device capability to achieve benchmarked training outcomes as part of each ATO's course(s) approval and not just FSTD qualification level
  - ICAO 9625 Edition 4 update to add EBT competencies in its training vs FSTD features/fidelity matrix

17

## TAKEAWAYS

- **Todays FPTs / FTDs are hugely more capable, affordable and faithful in fidelity on a par with the Level D FFS and the actual aircraft than ever before**
- **RAeS Spring 2017 Simulation Group Conference / CAT Magazine APATS Issue**

  **"Training / FSTD regulatory framework MUST evolve to... "*better recognise FPT/FTD current capabilities and find a way to benchmark their true fidelity, standardising characteristics and performance, and defining appropriate purposes and uses for the various levels of devices as we continue to evolve our training methodologies*"**

18

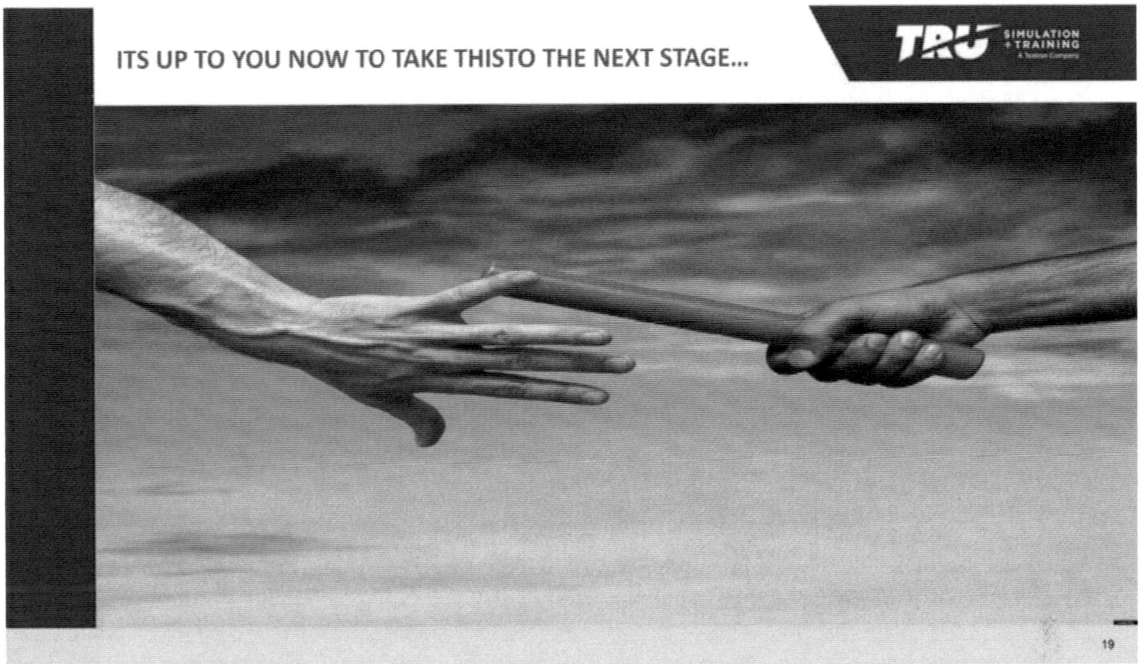
ITS UP TO YOU NOW TO TAKE THISTO THE NEXT STAGE...
TRU SIMULATION + TRAINING
19

TRANSFORM YOUR REALITY
THANK YOU
TRU SIMULATION + TRAINING
A Textron Company

SATCE - Emerging FSTD Technology
Results from Initial Trials
Dr Jeremy Goodman MRAeS
Senior Manager, Product Development
QUADRANT GROUP
apats2017
SINGAPORE
ASIA PACIFIC
AIRLINE TRAINING SYMPOSIUM
29-30 August 2017, Singapore

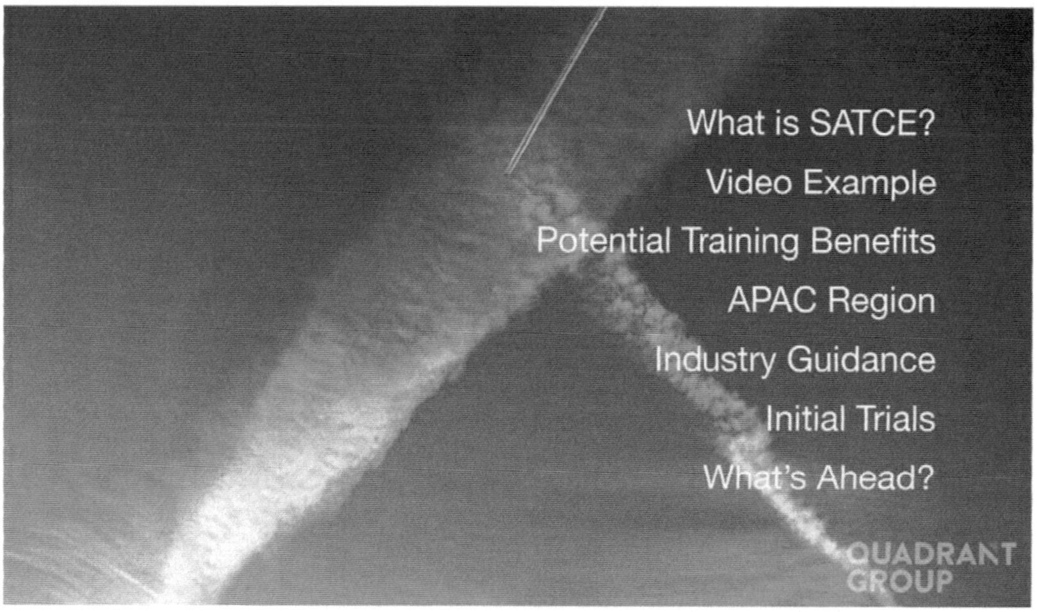
What is SATCE?
Video Example
Potential Training Benefits
APAC Region
Industry Guidance
Initial Trials
What's Ahead?
QUADRANT GROUP

# Simulated ATC Environment (SATCE)

"The simulation of other traffic entities within an airspace or ground environment, along with the associated ATC radio and data communications to other traffic and the ownship."

ICAO Doc 9625 Edition 4 (2015)

QUADRANT GROUP

"The simulation of **other traffic** entities within an airspace or ground environment, along with the associated **ATC radio and data communications** to other traffic and the ownship."

ICAO Doc 9625 Edition 4 (2015)

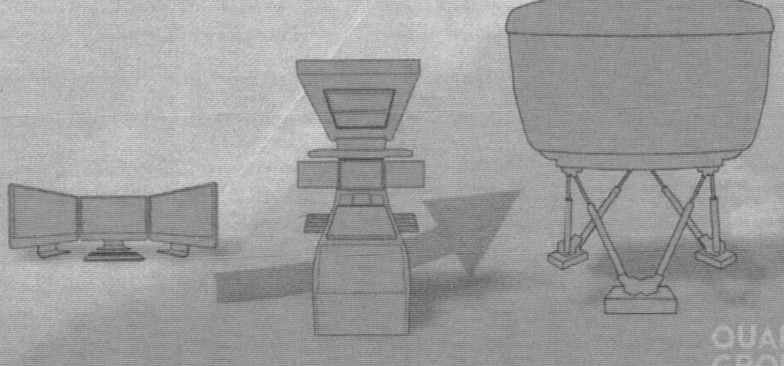

QUADRANT GROUP

# Video Example

- Straight-in Approach Scenario in to LHR, with a Go-Around
- Quadrant Group's implementation of SATCE, called INTERACT
- Integrated proof-of-concept, mid-2016
- FFS B737 Classic
- Professional Crew (TRE/TRI)

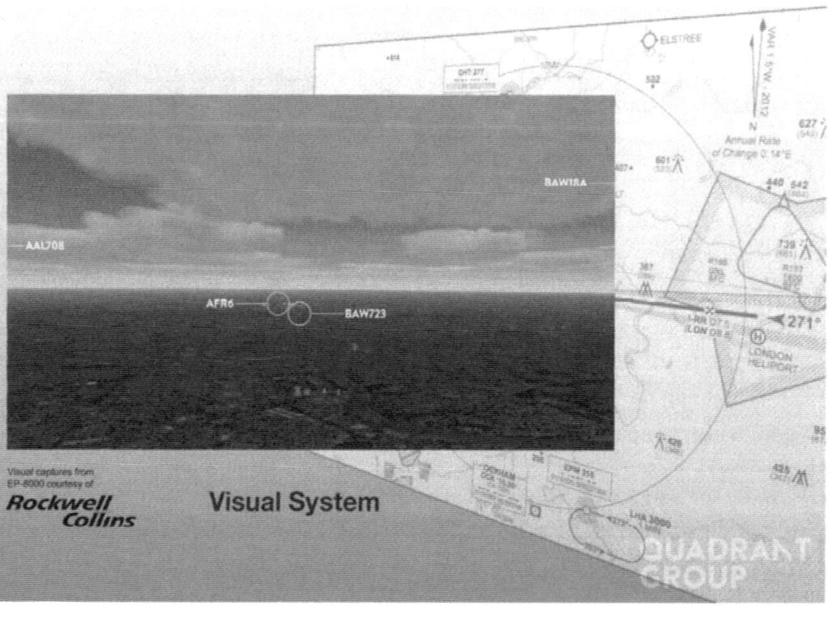


Visual captures from EP-8000 courtesy of

Rockwell Collins

Visual System

QUADRANT GROUP

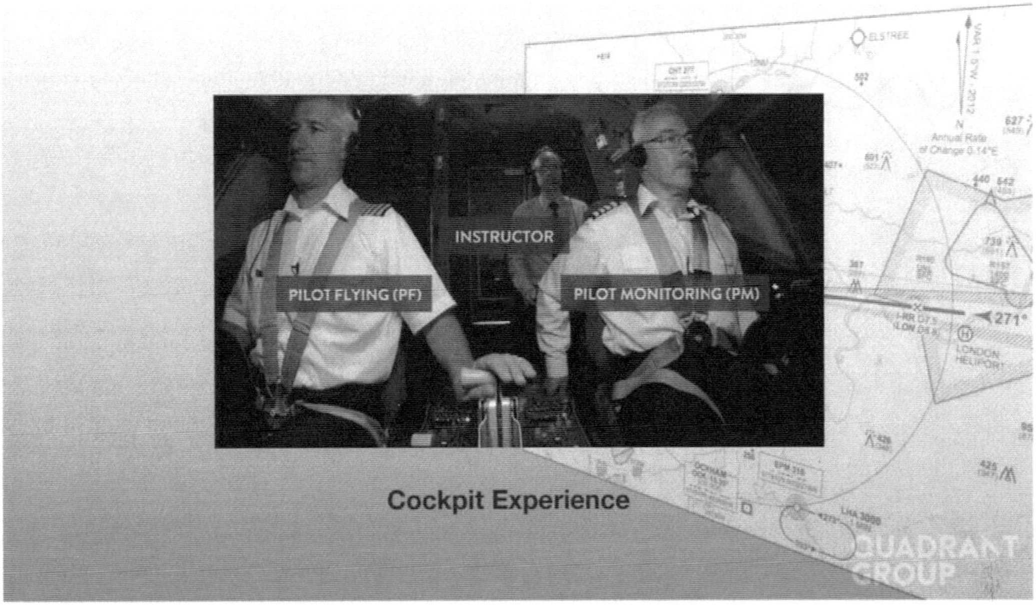
INSTRUCTOR
PILOT FLYING (PF)
PILOT MONITORING (PM)
271°
LONDON HELIPORT
Cockpit Experience
QUADRANT GROUP

Potential Training Benefits
QUADRANT GROUP

## Training Scenarios

**Enhanced Training Scenarios:**

- Runway Incursions
- Go Arounds, Missed Approaches
- Rejected Takeoff
- Low Visibility Procedures
- Loss of Separation / TCAS Events

**Just-in-time Training:**

- Flight Rehearsal
- Airport / Airspace / Familiarity

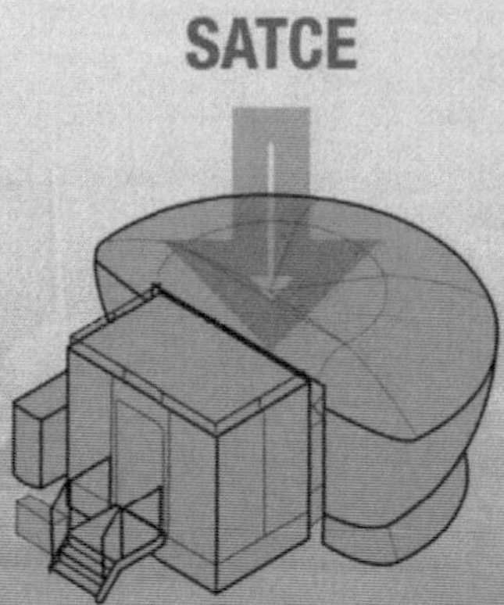


QUADRANT GROUP

## Competencies

**SATCE** supports **competency-based training** by providing a **higher-fidelity experience** of the real-world operational environment.

**EBT | MPL | CRM | TEM | HF | NTS**

QUADRANT GROUP

Competencies
Manual of
Evidence-based Training
ICAO Doc 9995
Manual of Evidence-based Training
1st Edition - 2013
QUADRANT
GROUP

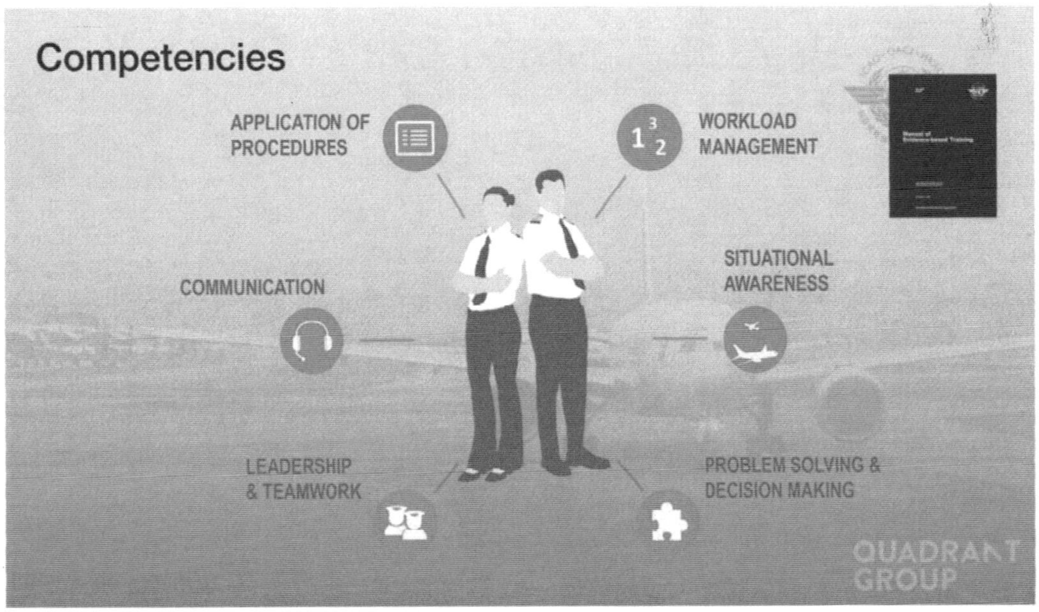
Competencies
APPLICATION OF PROCEDURES
WORKLOAD MANAGEMENT
COMMUNICATION
SITUATIONAL AWARENESS
LEADERSHIP & TEAMWORK
PROBLEM SOLVING & DECISION MAKING
QUADRANT
GROUP

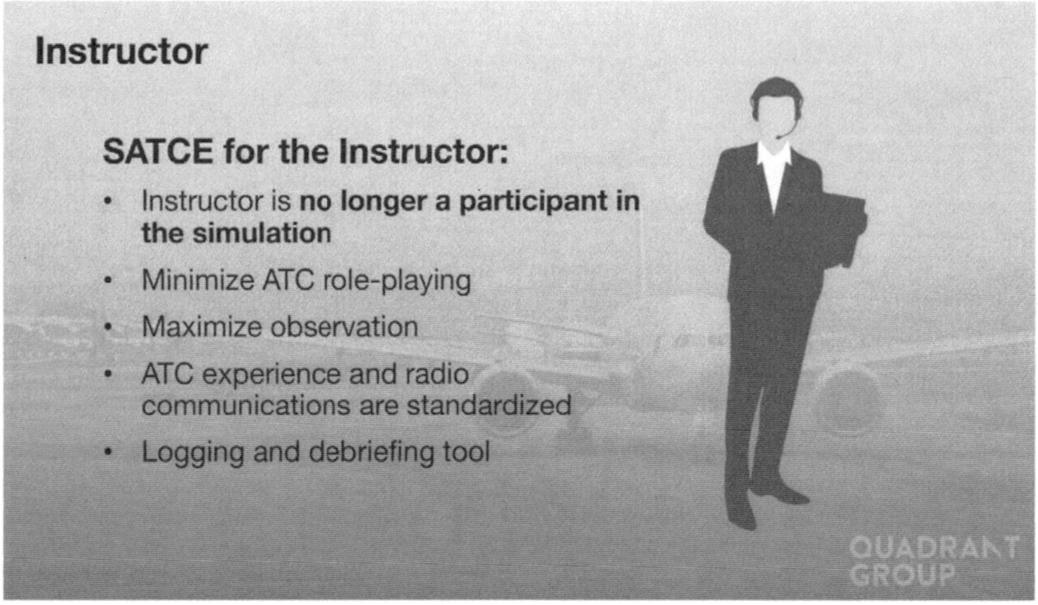
Instructor
SATCE for the Instructor:
• Instructor is **no longer a participant in the simulation**
• Minimize ATC role-playing
• Maximize observation
• ATC experience and radio communications are standardized
• Logging and debriefing tool
QUADRANT GROUP

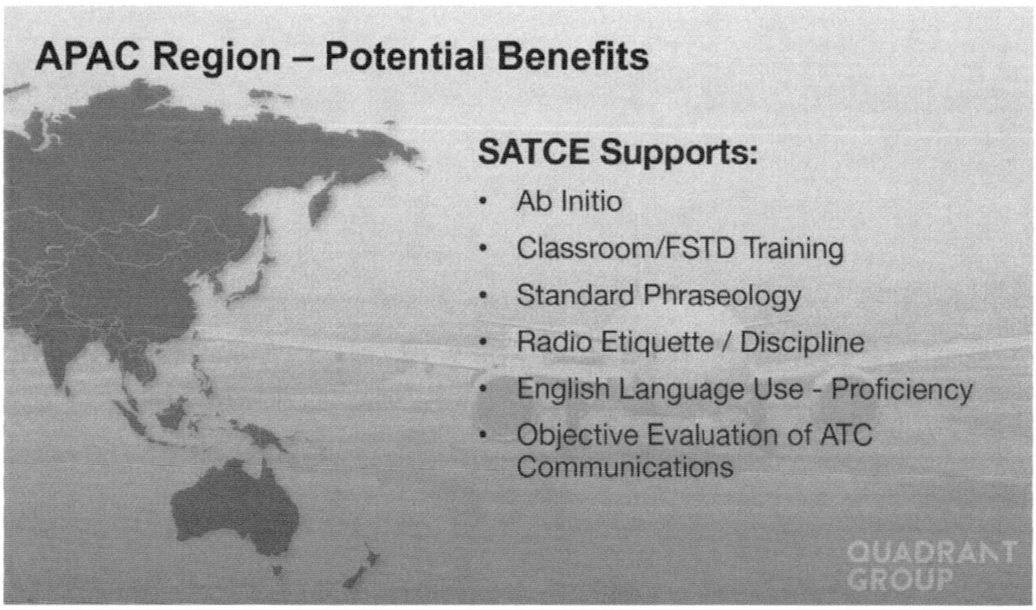
APAC Region – Potential Benefits
SATCE Supports:
• Ab Initio
• Classroom/FSTD Training
• Standard Phraseology
• Radio Etiquette / Discipline
• English Language Use - Proficiency
• Objective Evaluation of ATC Communications
QUADRANT GROUP

## Industry Guidance

- Emerging, maturing industry guidance on SATCE
- Required by EASA for Part-FCL for the MPL (2011)
- Recommended in:
  - ICAO Doc 9625 4th Ed. FSTD
  - ICAO Doc 9868 PANS-TRG
  - IATA / IFALPA 2nd Ed. Guidance for MPL
  - ARINC Spec 439A

QUADRANT GROUP

## ARINC Spec 439A (2016)

- Most comprehensive guidance
- SATCE recommended for **all levels** of training and device
- Includes guidance on **features, technology, maintenance** and **training approval / device qualification**
- Used to inform ICAO & other industry guidance

QUADRANT GROUP

ARINC Spec 439A (2016)
Facilitated by the
FSEMC Working Group on SATCE
ARINC
SIMULATED AIR TRAFFIC CONTROL ENVIRONMENTS IN FLIGHT SIMULATION TRAINING DEVICES
ARINC SPECIFICATION 439A
FlightSafety
FedEx
UFA
EASA
AIRBUS
THALES
ADACEL
QUADRANT GROUP
CAE
BOEING
Rockwell Collins
Airliner1
ROYAL AERONAUTICAL SOCIETY
SIMULATION + TRAINING
QUADRANT GROUP


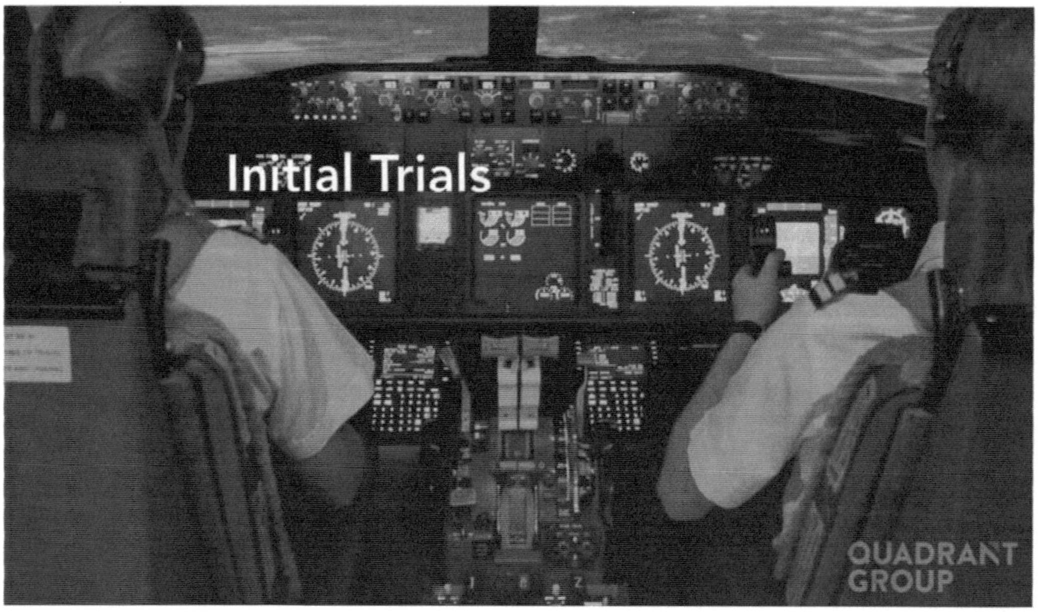

Initial Trials
QUADRANT GROUP

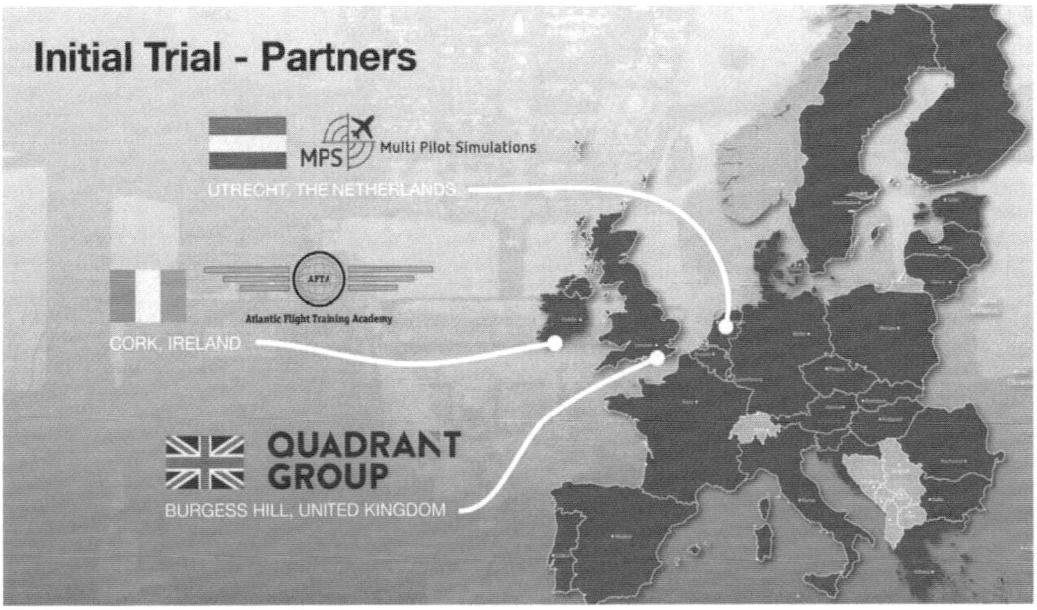
Initial Trial - Partners
MPS Multi Pilot Simulations
UTRECHT, THE NETHERLANDS
Atlantic Flight Training Academy
CORK, IRELAND
QUADRANT GROUP
BURGESS HILL, UNITED KINGDOM

Trial Device
MPS Multi Pilot Simulations
B737 NG FNPT-II
RSi

## Trial Objectives

QUADRANT
INTERACT™

1. Integrate QUADRANT INTERACT™
2. Test Technologies:
   - Speech Recognition (SR)
   - Artificial Intelligence (AI)
3. Mature Features
4. Improve Usability
5. Validate Training Benefits

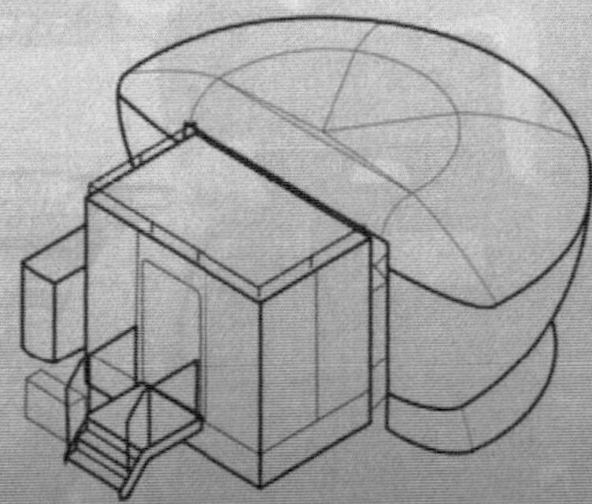

QUADRANT GROUP
Atlantic Flight Training Academy
MPS Multi Pilot Simulations
QUADRANT GROUP

## Trial Progress

QUADRANT GROUP

- ✓ Phases 2/3 Underway
- ✓ Significant Integration Completed
- ✓ Experienced Instructors Involved
- ✓ Multiple Training Scenarios
- ✓ Technology Showing Promise
- ✓ Next Steps – wider exposure during MCC / JOC training

AFTA
Atlantic Flight Training Academy
MPS Multi Pilot Simulations
QUADRANT GROUP

What's Ahead?
• Ongoing Trials - validate value added
• Initial Sales
• Early Adopters
• Maturing Guidance / Requirements
QUADRANT GROUP

Thank you
QUADRANT GROUP
Dr Jeremy Goodman MRAeS
Senior Manager, Product Development

# AUTHOR INDEX